Physics, Mathematics, and All that Quantum Jazz

Kinki University Series on Quantum Computing

Editor-in-Chief: Mikio Nakahara *(Kinki University, Japan)*

ISSN: 1793-7299

Published

Vol. 1 Mathematical Aspects of Quantum Computing 2007
edited by Mikio Nakahara, Robabeh Rahimi (Kinki Univ., Japan) &
Akira SaiToh (Osaka Univ., Japan)

Vol. 2 Molecular Realizations of Quantum Computing 2007
edited by Mikio Nakahara, Yukihiro Ota, Robabeh Rahimi, Yasushi Kondo &
Masahito Tada-Umezaki (Kinki Univ., Japan)

Vol. 3 Decoherence Suppression in Quantum Systems 2008
edited by Mikio Nakahara, Robabeh Rahimi & Akira SaiToh
(Kinki Univ., Japan)

Vol. 4 Frontiers in Quantum Information Research: Decoherence, Entanglement,
Entropy, MPS and DMRG 2009
edited by Mikio Nakahara (Kinki Univ., Japan) &
Shu Tanaka (Univ. of Tokyo, Japan)

Vol. 5 Diversities in Quantum Computation and Quantum Information
edited by edited by Mikio Nakahara, Yidun Wan & Yoshitaka Sasaki
(Kinki Univ., Japan)

Vol. 6 Quantum Information and Quantum Computing
edited by Mikio Nakahara & Yoshitaka Sasaki (Kinki Univ., Japan)

Vol. 7 Interface Between Quantum Information and Statistical Physics
edited by Mikio Nakahara (Kinki Univ., Japan) &
Shu Tanaka (Univ. of Tokyo, Japan)

Vol. 8 Lectures on Quantum Computing, Thermodynamics and Statistical Physics
edited by Mikio Nakahara (Kinki Univ., Japan) &
Shu Tanaka (Univ. of Tokyo, Japan)

Vol. 9 Physics, Mathematics, and All That Quantum Jazz
edited by Shu Tanaka (Univ. of Tokyo, Japan), Masamitsu Bando (Kinki
Univ., Japan) & Utkan Güngördü (Kinki Univ., Japan)

Kinki University Series on Quantum Computing – Vol. 9

editors

Shu Tanaka
University of Tokyo, Japan

Masamitsu Bando
Kinki University, Japan

Utkan Güngördü
Kinki University, Japan

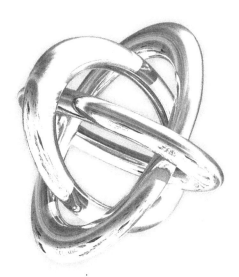

Physics, Mathematics, and All that Quantum Jazz

World Scientific

NEW JERSEY · LONDON · SINGAPORE · BEIJING · SHANGHAI · HONG KONG · TAIPEI · CHENNAI

Published by

World Scientific Publishing Co. Pte. Ltd.
5 Toh Tuck Link, Singapore 596224
USA office: 27 Warren Street, Suite 401-402, Hackensack, NJ 07601
UK office: 57 Shelton Street, Covent Garden, London WC2H 9HE

British Library Cataloguing-in-Publication Data
A catalogue record for this book is available from the British Library.

Kinki University Series on Quantum Computing — Vol. 9
PHYSICS, MATHEMATICS, AND ALL THAT QUANTUM JAZZ

Copyright © 2014 by World Scientific Publishing Co. Pte. Ltd.

ISBN 978-981-4602-36-5

In-house Editor: Rhaimie Wahap

Printed in Singapore

PREFACE

This volume contains articles presented at the Summer Workshop on Physics, Mathematics, and All That Quantum Jazz, held from 7 to 9 August, 2013 at Kinki University, Osaka, Japan. One of the purposes of the symposium was to exchange and share ideas among researchers working in various fields related to quantum physics. Speakers were asked to make their presentations accessible to other researchers with diverse backgrounds. Articles in this volume reflect this request and they should be accessible not only to forefront researchers but also to beginning graduate students or advanced undergraduate students. This symposium was also held in honor of Professor Mikio Nakahara, celebrating his 60th birthday.

This symposium was supported financially by the Interdisciplinary Graduate School of Science and Engineering, Kinki University under the project "Research Center for Quantum Computing".

We would like to thank all the lecturers and participants, who made this symposium invaluable. Finally, we would like to thank Rhaimie B Wahap of World Scientific for her excellent editorial work.

Shu Tanaka
Masamitsu Bando
Utkan Güngördü

Tokyo and Osaka, October 2013

CONTENTS

viii

Summer Workshop on
Physics, Mathematics, and All That Quantum Jazz

Kinki University (Osaka, Japan)

---

7 – 9   August   2013

---

7 August 2013

[Chair: Utkan Güngördü (Kinki University)]
Shu Tanaka (The University of Tokyo)
> Opening

Yukihiro Ota (RIKEN)
> Implementing Measurement Operations in Linear Optical and Solid-State Qubits

Akira SaiToh (National Institute of Informatics)
> Fast and Accurate Simulation of Quantum Computing by Multi-Precision MPS: Recent Development

[Chair: Keisuke Fujii (Kyoto University)]
Shu Tanaka (The University of Tokyo)
> Entanglement Properties in Two-Dimensional Quantum Systems

Yutaka Shikano (Institute for Molecular Science)
> On the Signal Amplification – From Weak-Value Amplification –

[Chair: Toshihiro Sato (RIKEN)]
Keisuke Fujii (Kyoto University)
> Topological Protection of Quantum Information

Yuya Seki (Tokyo Institute of Technology)
> Quantum Annealing with Antiferromagnetic Fluctuations

Ryo Tamura (National Institute of Materials Science)
> Toward An Alternative Method to Quantum Annealing – Quest for New Type of Fluctuation –

Masahito Tada-Umezaki (University of Toyama)
Computational Analysis of the First Stage of the Photosynthetic System, The Light-Dependent Reaction, by Quantum Chemical Simulation Method

8 August 2013

[Chair: Masamitsu Bando (Kinki University)]
Elham Hosseini (Osaka City University)
Two-Qubit Gate Operation Applied on Nearest Neighboring Qubits in a Neutral Atom Quantum Computer

Chiara Bagnasco (Kinki University)
A Simple Operator Quantum Error Correction Scheme Avoiding Fully Correlated Errors

Akihiro Ishibashi (Kinki University)
Black Hole Predictability, Classical and Quantum

[Chair: Ryo Tamura (National Institute of Materials Science)]
Toshihiro Sato (RIKEN)
Classical Field Simulation of Finite-Temperature Bose Gases

Kenichi Kasamatsu (Kinki University)
Atomic Quantum Simulations of Lattice Gauge Theory: Effect of Gauge Symmetry Breaking

[Chair: Yukihiro Ota (RIKEN)]
Yiu-Tung Poon (Iowa State University)
Quantum Error Correction for Collective Rotation Channels on Qudits

Nung-Sing Sze (The Hong Kong Polytechnic University)
Quantum Error Correction of Phase-Flip Channel

Mikio Nakahara (Kinki University)
My Life As a Quantum Physicist

9 August 2013

[Chair: Akira SaiToh (National Institute of Informatics)]
Chi-Kwong Li (College of William & Mary)
    Quantum States with the Same Reduced States

Utkan Güngördü (Kinki University)
    Non-Adiabatic Holonomic Quantum Computation

Masamitsu Bando (Kinki University)
    Composite Quantum Gates for Precise Quantum Control

[Chair: Shu Tanaka (The University of Tokyo)]
Sho Sugiura (The University of Tokyo)
    New Formulation of Statistical Physics Using Thermal Pure Quantum States

Tatsuhiko Ikeda (The University of Tokyo)
    Thermodynamics in Unitary Time Evolution

Takahiro Sagawa (The University of Tokyo)
    Second Law of Thermodynamics with the QC-Mutual Information

Shu Tanaka (The University of Tokyo)
    Closing

LIST OF PARTICIPANTS

Bagnasco, Chiara	Kinki University, Japan
Bando, Masamitsu	Kinki University, Japan
Fujii, Keisuke	Kyoto University, Japan
Güngördü, Utkan	Kinki University, Japan
Hosseini Lapasar, Elham	Osaka City University, Japan
Ikeda, Tatsuhiko	The University of Tokyo, Japan
Imai, Masayuki	Kinki University, Japan
Izumi, Shuzo	Kinki University, Japan
Kasamatsu, Kenichi	Kinki University, Japan
Kojima, Shoko	Kinki University, Japan
Li, Chi-Kwong	College of William & Mary, USA
Miyoshi, Hiroyuki	Kyoto Sangyo University, Japan
Morimoto, Tohru	Doshisha University, Japan
Nakagawa, Nobuo	Kinki University, Japan
Nakahara, Mikio	Kinki University, Japan
Ohno, Yasuo	Kinki University, Japan
Ohzeki, Masayuki	Kyoto University, Japan
Ota, Yukihiro	RIKEN, Japan
Poon, Yiu-Tung	Iowa State University, USA
Sagawa, Takahiro	The University of Tokyo, Japan
Sato, Toshihiro	RIKEN, Japan
Seki, Yuya	Tokyo Institute of Technology, Japan
Shikano, Yutaka	Institute for Molecular Science, Japan
Sugiura, Sho	The University of Tokyo, Japan
Sze, Raymond	The Hong Kong Polytechnic University, Hong Kong
Tada-Umezaki, Masahiro	University of Toyama, Japan
Tamura, Ryo	National Institute for Materials Science, Japan
Tanaka, Shu	The University of Tokyo, Japan
Yamagata, Koichi	Osaka University, Japan

MY LIFE AS A QUANTUM PHYSICIST

MIKIO NAKAHARA

Research Center for Quantum Computing
Interdisciplinary Graduate School of Science and Engineering
Kinki University
Higashi-Osaka, 577-8502, Japan
E-mail: nakahara@math.kindai.ac.jp
http://alice.math.kindai.ac.jp

I will summarize my research carrier in the past 36 years and present future prospects of my research interest.

Keywords: Neutron Star/Pulsar, Superfluid ^{3}He, Superconductivity, Polyacetylene, Supersymmetry, Bose-Einstein Condensate, Quantum Information, Quantum Computing, Quantum Control

1. My Research Career to Date

I entered Kyoto University[1] in 1971, where I studied physics, mathematics, chemistry and astronomy. There were no departments for undergraduate students and we were free to study any subjects we were interested in. I have learnt some parts of my mathematical background from pure mathematicians. I was lucky enough to learn differential geometry and Lie groups from such experts as Tohru Morimoto, one of the participants, and Makoto Matsumoto in my undergraduate student days.

When I was in the fourth grade, I joined Chushiro Hayashi's group[2] to conduct undergraduate research project. I have chosen to study neutron stars, which was a candidate of a pulsar (PSR B1919+21) discovered in 1967.[3] I first solved the Einstein equation simultaneously with the equation of state of a free neutron gas numerically, using a Hitachi computer on campus, to find the density profile of a neutron star. This is essentially a repetition of a work done by Oppenheimer and Volkoff in 1939 using a mechanical calculator.[4] I looked at several papers analyzing the critical mass M_c of a neutron star, beyond which a black hole is formed, and learned that it depends on the interaction between neutrons. I tried to include the neutron-neutron interaction potential in the equation of state using the Brückner theory of many body quantum system with a hard core potential[5,6] with the Hamada-Johnston nuclear potential,[7] which was widely used at that time. I prepared a FORTRAN program, which solved the set of differential equations by the Runge-Kutta method, with more than 1,000 cards.[a] Nevertheless, the most powerful computer in Japan at that time did not output the results within a CPU time[b] assigned to a student.

Then I joined Toshihiko Tsuneto's group as a postgraduate student. At that time, he was working on neutron star cooling by neutrino emission with Naoki Itoh.[8] He was also working on $^3P_2$ superfluidity of neutrons,[9] which was supposed to exist in the interior of a neutron star. I was expecting to work on these subjects when I joined the research group.

It was a few years after some peculiar behavior of ^{3}He at extremely low temperature of $\sim$ mK was found. It became clear then that this strange behavior was attributed to the superfluidity of ^{3}He. I decided to do research on rotating superfluid ^{3}He-A under the supervision of Toshihiko Tsuneto, Tetsuo Ohmi and Toshimitsu Fujita, the results of which were published as several papers and became the subject of my MSc thesis.[10] In these works, we analyzed the vortex lattice structures that may appear when the superfluid is rotated. The unit vortex in this lattice is different from the one that appears in superfluid ^{4}He or ordinary superconductors in that it has no singular core. Mathematically, this is due to the well known fact of the fundamental group $\pi_1(SO(3)) = \mathbb{Z}_2$.[11] Here the algebra is defined by $0 + 0 = 1 + 1 = 0$ and $0 + 1 = 1 + 0 = 1$. This implies that two singluar votices add up to form a no vortex! Half of my MSc thesis was devoted to the application of homotopy theory for classification of defects in condensed matter system and this is how I was interested in topology. The above homotopy formula has been repeatedly used in my work in different contexts.

When I was in the third year of the PhD program, I found there were more jobless senior members than lower grade graduate students in the group. To have more chances to get a job, I asked Prof. Tsuneto to send me to the US. He immediately recommended me to continue research at University of Southern California[12] with Kazumi Maki. I joined Kazumi's group in August 1980. Professor Hideki Shirakawa and his collaborators

---

[a] Younger people may not know how a computer program was prepared in the '70s. Each program line was represented by a card with punched holes. You might imagine the disaster taking place when one drops a deck of 1,000 cards!

[b] I vaguely remember that it was on the order of 20 minutes.

reported that doping polyacetylene with iodine extremely enhanced the conductivity of polyacetylene.[13] We were working on the properties of solitons in polyacetylene. The electrons in a long chain of polyacetylene is described by a $1+1$-dimensional Dirac Hamiltonian with a background potential corresponding to the lattice deformation. By connecting two chains with opposite sense of deformation, an intermediate region is formed at the boundary, which is called a soliton. This soliton has a remarkable property in that its energy eigenstate appears in the exact center of the band gap, whose existence is mathematically guaranteed by a type of an index theorem. I did my PhD work by evaluating the quantum corrections of the soliton mass due to the background lattice vibration (phonons).[14] I also continued research in superfluid ^{3}He and published a few papers on this subject. During this period, I notified that Finnish group (ROTA) was working on rotating superfluid ^{3}He. I could not predict, however, that I would visit Finland very often 20 years later.

After a hard work at USC, I wanted to study more mathematical subjects and decided to spend one year as a student at Mathematics Department, King's College, University of London.[15] The program I was enrolled was called "Classical and Quantum Gravity". There, I had a plenty of time to acquire new background and to talk to people from totally different disciplines. Many of my friends thought spending one year as a student after getting PhD could be a waste of time (and also money) but I believe the year I spent in King's have enriched my way of thinking and research immeasurably. When I was in King's I was interested in supersymmetry at finite temperature. Supersymmetry is a symmetry between Bosons and Fermions.[16] Since they obey different statistics at finite temperature, it was natural to expect that supersymmetry was broken spontaneously at finite temperature. Although there were a few papers on supersymmetry at finite temperature already, none of them seemed to be reasonable enough to me.

After I finished the program at King's, I moved to University of Alberta,[17] Canada as a postdoc working for Hiroomi Umezawa.[18] I continued research on supersymmetry at finite temperatures with Hiroomi Umezawa, Hideki Matsumoto, Yoshimasa Nakano and Noboru Yamamoto, which resulted in several papers on this subject.[19-23] There I learned finite temperature field theory (Thermo Field Dynamics[24]), which is relevant to study symmetry involving time. Conventional Matsubara formalism, in which the time axis is compactified to a circle is not very much useful for analyzing supersymmetry and Poincare symmetry since time is compactified form the beginning. I also learned many techniques in relativistic field theory from my friend Yoshimasa Nakano, who is now a professor at Kinki University Kyushu campus. We mainly analyzed the properties of the Goldstone fermions associated with the finite temperature phase transition.

While I was working in Alberta, I met several visitors and I collaborated with them from time to time. Masashi Tachiki visited us in one summer and we did work on the tunneling spectrum of the anisotropic superconductors,[25] candidates of heavy fermion superconductors actively studied at that time. I was also attracted by a paper by Ho, Fulco, Schrieffer and Wilczek on the relevance of charge fractionization in the domain wall of superfluid ^{3}He.[26] I applied their analysis to an exact solution of a domain wall in a thin film of superfluid ^{3}He and got more or less closed form of the exact solution.[27] The only problem is that the exact solution was obtained for an unrealistic large paramagnon parameter but it should still reflect the general behavior for realistic paramagnon parameters.

In 1985, I moved to University of Sussex, UK[28] to be a postdoc working under a project on applications of quantum field theory to condensed matter physics, directed by David Bailin and Norman Dombey. I continued research on superfluid ^{3}He and polyacetylene with David Waxman, currently a biology/mathematics professor at Fudan University, China. At that time, I was looking for a tenure-track or tenured position. Some position openings required teaching experience but teaching was not a part of my postdoc duties. So I decided to give voluntary lectures on geometry and topology to postgraduate students and faculty members in condensed matter physics, particle theory and gravitational physics. I planned to give 20 lectures but it was interrupted by an offer of associate professorship from Shizuoka University, Japan.[29] I felt sorry to the audience to quit my lecture after 10 or 12 lectures. Then David Bailin suggested me to write a book based on my lecture note. He also talked to Douglas Brewer, the Dean of the School of Mathematics and Physics at that time and, at the same time, the Editor of "Graduate Student Series in Physics", published from Adam-Hilger, the book division of the Institute of Physics, UK, to include my future book in his series. I left sussex to take up my position in Japan in March 1986 and the book "Geometry, Topology and Physics"[11] was published in January 1990. It is amusing to know that the book has been used as a textbook/reference book in many universities worldwide, including King's and Sussex. I really thank David Bailin and also audience of my lectures for making the publication of this book possible.

After I assumed the associate professorship in Shizuoka, I spent busy days with heavy teaching and administration duties and had not much time to do research. I applied for JSPS (Japan Society for Promotion of Science) visiting professorship and it was accepted. I spent one year at Sussex from 1989 to 1990 and worked with David Waxman, his postdoc Gareth Williams and student David Pattarini. We worked together on topological aspects of extended objects in superfluid ^{3}He and solitions in polyacetylene, which resulted in several publications.[30–33] Later I accepted two postdocs from Sussex, Gareth Williams and Andrew Poon, in Shizuoka. They were supported financially by JSPS (Japan Society for Promotion of Science).

I moved to Kinki University[34] in April 1993. I had and still have many collaborators in Kinki area and I thought it was better to move to this area for more active research. I also wanted to continue close collaboration with Tetsuo Ohmi. Right after I moved to Kinki, I started working on quantum tunneling. There was Grant-in-Aid for Scientific Research dedicated to research on quantum tunneling and I worked on tunneling of a metastable polyacetylene chain toward stable polyacetylene chain by the formation of an island of a stable polyacetylene.[35,36]

Bose-Einstein condensation (BEC) of a cold atomic gas has been predicted many decades ago. It was demonstrated physically in 1996 by two groups led by Eric Cornell and Wolfgang Ketterle, respectively. The condensate has rich internal structure since the associated hyperfine spin belongs to higher-dimensional irreducible representation of $SU(2)$. It was pointed out by Tetsuo Ohmi, Kazushige Machida and Tin-Lun Ho that the internal degrees of freedom play remarkable roles in the behavior of the BEC.[37,38] When I attended a Japanese Physical Society meeting around 1998 or 1999, I was asked to have discussion with Ohmi and Machida. It was about creating a vortex in BEC by simply manipulating the confining magnetic field. It seemed terribly simple and to have deep mathematical significance in their idea. I noticed that this could be another manifestation of the formula $\pi_1(SO(3)) = \mathbb{Z}$ and the algebra $0 = 1 + 1$ suggested that one could create a vortex of higher winding number out of a vortex-free order parameter. Later, it was pointed out that formation of a vortex with higher winding number was also regarded as a manifestation of imprinting Berry phase in the condensate by manipulating the magnetic field adiabatically. This "topological formation of vortices in BEC" has been a research subject over a decade, which resulted in several papers involving Japanese and Finnish groups.[39–44]

I stayed in Martti Salomaa's group in Helsinki University of Technology, Finland (currently known as Aalto University[45]) for 6 months from September 2001 till March 2002. I worked with Tetsuo Ohmi, Mikko Möttönen, a

PhD student then, Shinichiro Ogawa, Tomoya Isoshima and Hisanori Shimada on topological formation of a vortex in BEC of alkali atoms. During my stay in Helsinki, I was asked by Martti to lecture on quantum computing. This is how I started working on quantum information and quantum computing. While I was staying in Helsinki, I found several excellent PhD students, such as Mikko Möttönen, Juha Vartianen, Antti Niskanen, among others. I started collaboration with Martti and these students. I visited Helsinki a few times each year and we did several works on holonomic quantum gate implementation.[46,47] We were also interested in actual implementation of Shor's algorithm assuming some particular Hamiltonian and found that it required $\sim 10^5$ gates and 21 qubits to have factorization $21 = 3 \times 7$.[48]

I started several projects with Yasushi Kondo on NMR quantum computing in the beginning of the 21st century. We demonstrated implementation of simple quantum algorithms, implementation of high precision gates, acceleration of quantum algorithms using Cartan decomposition of SU(4) and many other subjects. Implementation of high-precision gates is still an active research subject also involving Masamitsu Bando, Tsubasa Ichikawa and Utkan Güngördü.

Chi-Kwong Li once sent me an email. I vaguely remember that he was using a textbook on quantum computing I wrote with Tetsuo Ohmi and his email was comments or questions about our book. Then I had a chance to visit him at The College of William & Mary and gave a general introduction to quantum computing for mathematics students. Then I visited his home country, Hong Kong, to attend a workshop Chi-Kwong and his collaborators, Yiu-Tung Poon and Raymond Nung-Sing Sze organized. I was with one of my collaborators, Hiroyuki Tomita, and it was the time we have worked out a variant of quantum error correcting code, which does not require ancillas for error syndrome readout. We were very much excited about it and we started collaboration since then. We published several works on quantum error correcting codes for fully correlated noise. Later, my PhD student, Utkan Güngördü joined this project.

My recent research activities were/will be introduced my collaborators and students and let me quit on my past research subjects here.

2. Future Prospects

I am currently involved in Grant-in Aid for Scientific Research on Innovative Areas from the Ministry of Education, Culture, Sports, Science and Technology (MEXT). The title of the project is "Topological Quantum

Phenomena" and I am working on the half-quantum vortex (HQV) with Tetsuo Ohmi under this scheme.

Recently, I sent a proposal of a new book to a publisher. This will be written with Kaiki Inoue and Akihiro Ishibashi with the title "Geometry and Topology of Spacetime". Space is still the final frontier.

I would like to apologize to collaborators whose names/works are not mentioned here. I simply have no time to complete the full list of my research.

Acknowledgments

I would like to thank the organizers of this workshop, who made this fantastic meeting possible. 60th birthday is special in Chinese and Japanese culture. Our calender is defined modulo 60 and 60 years after one's birth, one's age is reset to zero. This phenomena is known as KANREKI in Japanese. I hope all of us are healthy till my next KANREKI and meet here again in the year 2072.

References

1. http://www.kyoto-u.ac.jp/en
2. http://en.wikipedia.org/wiki/Chushiro_Hayashi
3. http://en.wikipedia.org/wiki/PSR_B1919%2B21
4. J. R. Oppenheimer and G. M. Volkoff, Phys. Rev. **55**, 374 (1939).
5. K. A. Brueckner, C. A. Levinson and H. M. Mahmoud, Phys. Rev. **95**, 217 (1954). K. A. Brueckner, J. L. Gammel and J. T. Kubis, Phys. Rev. **118**,1095 (1960).
6. B. D. Day, Rev. Mod. Phys. **39**, 719 (1967).
7. T. Hamada and I. D. Johnston, Nucl. Phys. **34**, 382 (1962).
8. N. Itoh and T. Tsuneto, Prog.Theor. Phys. **48**, 1849 (1972).
9. T. Fujita and T. Tsuneto, Prog. Theor. Phys. **48**, 766 (1972).
10. http://repository.kulib.kyoto-u.ac.jp/dspace/handle/2433/89528
11. M. Nakahara, *Geometry, Topology and Physics* 2nd ed. (Taylor & Francis, 2003).
12. http://www.usc.edu/
13. H. Shirakawa, E. J. Louis, A. G. MacDiarmid, C. K. Chiang and A. J. Heeger, J. Chem. Soc., Chem. Commun., 578 (1977).
14. Mikio Nakahara and Kazumi Maki, Phys. Rev. B **25**, 7789 (1982).
15. http://www.kcl.ac.uk/index.aspx
16. http://en.wikipedia.org/wiki/Supersymmetry
17. http://www.ualberta.ca/
18. http://en.wikipedia.org/wiki/Hiroomi_Umezawa
19. H. Matsumoto, M. Nakahara, Y. Nakano and H. Umezawa, Phys. Lett. **140** B, 53 (1984).

12

20. H. Matsumoto, M. Nakahara, Y. Nakano and H. Umezawa, Phys. Rev. D **29**, 2838 (1984).

21. M. Nakahara, Phys. Lett. **142** B, 395 (1984).

22. H. Matsumoto, M. Nakahara, Y. Nakano and H. Umezawa, Physica **15** D, 163 (1985).

23. H. Matsumoto, M. Nakahara, H. Umezawa and N. Yamamoto, Phys. Rev. D **33**, 2851 (1986).

24. H. Umezawa, H. Matsumoto and M. Tachiki, *Thermo field dynamics and condensed states*, (North-Holland, 1982).

25. M. Tachiki, M. Nakahara and R. Teshima, J. Mag. Mag. Mater., **52**, 161 (1985).

26. T. L. Ho, J. R. Fulco, J. R. Schrieffer and F. Wilczek, Phys. Rev. Lett. **52**, 1524 (1984).

27. M. Nakahara, J. Phys. C: Solid State Phys. **19**, L195 (1985).

28. http://www.sussex.ac.uk/

29. http://www.shizuoka.ac.jp/english/

30. M. Nakahara, D. Waxman and G. Williams, J. Phys. A: Math. Gen. **23**, 5017 (1990).

31. M. Nakahara and G. Williams, Prog. Theor. Phys. **86**, 315 (1991).

32. M. Nakahara, D. Waxman and G. Williams, J. Phys. C: Condens. Matter, **3**, 6743 (1991).

33. M. Nakahara, D. Waxman and G. Williams, Prog. Theor. Phys. **88**, 129 (1992).

34. http://www.kindai.ac.jp/english/

35. M. Nakahara and T. Ohmi, Prog. Theor. Phys. **94**, 311 (1995).

36. M. Nakahara, Phys. Lett. A **236**, 97 (1997).

37. T. Ohmi and K. Machida, J. Phys. Soc. Jpn., **67**, 1822 (1998).

38. T.-L. Ho, Phys. Rev. Lett. **81**, 742 (1998).

39. M. Nakahara, T. Isohima, K. Machid, S.-i. Ogawa and T. Ohmi, Physica B **284-288** (2000) 17 (2000).
    32. 33. 34.

40. T. Isoshima, M. Nakahara, T. Ohmi and K. Machida, Phys. Rev. A **61**, 063610 (2000).

41. S.-i. Ogawa, M. Möttönen, M. Nakahara, T. Ohmi and H. Shimada, Phys. Rev. A **66**, 013617 (2002).

42. M. Möttönen, N. Matsumoto, M. Nakahara and T. Ohmi, J. Phys.: Condens. Matter **14**, 13481 (2002).

43. Y. Kawaguchi, M. Nakahara and T. Ohmi, Phys. Rev. A **70**, 043605 (2004).

44. Y. Kawaguchi, M. Nakahara and T. Ohmi, J. Low Temp. Phys. **138**, 699 (2005).

45. http://www.aalto.fi/en/

46. A. O. Niskanen, M. Nakahara and M. M. Salomaa, Quantum Inf. Comput. **2**, 560 (2002).

47. A. O. Niskanen, M. Nakahara and M. M. Salomaa, Phys. Rev. A **67**, 012319 (2003).
48. J. J. Vartiainen, A. O. Niskanen, M. Nakahara and M. M. Salomaa, Phys. Rev. A **70**, 012319 (2004).

A REVIEW ON OPERATOR QUANTUM ERROR CORRECTION

DEDICATED TO PROFESSOR MIKIO NAKAHARA
ON THE OCCASION OF HIS 60TH BIRTHDAY

CHI-KWONG LI

Department of Mathematics, College of William & Mary, Williamsburg, VA 23187-8795, USA. Email: ckli@math.wm.edu

YIU-TUNG POON

Department of Mathematics, Iowa State University, Ames, IA 50011, USA. Email:ytpoon@iastate.edu

NUNG-SING SZE

Department of Applied Mathematics, The Hong Kong Polytechnic University, Hung Hom, Hong Kong. Email:raymond.sze@polyu.edu.hk

In this article, we first review some general mathematical framework for operator quantum error correction. Then implementation of these schemes on collective noise was discussed.

Keywords: Quantum error correction; Decoherence free subspace; Noiseless subsystem; Collective noise.

1. Introduction

Quantum information science concerns information theory that makes use of quantum nature of the microscopic world. In reality, quantum systems are vulnerable to disturbance from an external environment, which can lead to decoherence in the system. Thus, the system must be protected from the environmental noise to maintain the accuracy of the information stored in the quantum registers. In order to realize a working and dependable quantum computational device, researchers and engineers have to overcome this difficulty. One of the most well accepted candidates for overcoming decoherence is Quantum Error Correction (QEC).[12,19,31,32]

During the mid 1990s, Shor[35] and Steane[37] suggested examples on how data could be redundantly encoded in the states of a quantum system, and how the redundancy could be used to protect the data. Then this research topic took flight in a short period of time by many researchers, see Refs. 1,3,6,7,15–17,32,33,37,38 and their references therein, and is still under rapid development. Different approaches have been developed in the study of quantum error correction. For example, Calderbank, Rain, Shor, Sloane, and Gottesman considered the method of quantum stabilizer codes.[5–7,13] Another class of codes, called non-additive quantum error correcting codes,[22,36,43–45] was proposed and studied by other researchers. Further, experiments have been done to implement basic quantum error correction schemes.[2,9,11,18,34] In literatures, QEC schemes are roughly separated into two classes: one class employs extra ancilla qubits for error syndrome readout, see for example Refs. 30,31, while another class, called operator quantum error correction (OQEC), employs higher rank projection operators based on the Knill-Laflamme result; for example, see Refs. 17,21,32.

In this article, we will focus on operator quantum error correction. A brief review will be given in section 2 while the implementation of this scheme on collective noise will be presented in section 3.

2. General mathematical framework

Mathematically, quantum states ρ in quantum system are represented as density operators (trace one positive semidefinite operators) in $\mathcal{B}(\mathcal{H})$, where $\mathcal{H}$ is a finite dimensional complex Hilbert space (identified as $\mathbb{C}^n$). A quantum channel is a trace preserving completely positive (TPCP) linear map Φ on $\mathcal{B}(\mathcal{H})$ with the operator sum presentation:[8]

$$\Phi(\rho) = \sum_{j=1}^{r} E_j \rho E_j^{\dagger} \quad \text{with} \quad \sum_{j=1}^{r} E_j^{\dagger} E_j = \mathbf{1}, \tag{1}$$

where $E_1, \ldots, E_r \in \mathcal{B}(\mathcal{H})$ are known as the error operators of the quantum channel (they are also known as the Kraus-Choi operators). Recall that $\mathcal{H}$ and $\mathcal{B}(\mathcal{H})$ can be identified as the n-dimensional complex vector space $\mathbb{C}^n$ and the space of $n \times n$ complex matrices M_n respectively when $\mathcal{H}$ has finite dimension.

2.1. *Quantum error correcting code*

A quantum error correcting code (QECC) is a k-dimensional subspace $\mathcal{C}$ of $\mathcal{H}$ such that there is another quantum channel $\Psi : \mathcal{B}(\mathcal{H}) \to \mathcal{B}(\mathcal{H})$ with the property that $\Psi \circ \Phi(\rho) = \rho$ for all ρ satisfying $P_{\mathcal{C}} \rho P_{\mathcal{C}} = \rho$, where $P_{\mathcal{C}}$ is the orthogonal projection of $\mathcal{H}$ onto the subspace $\mathcal{C}$. The quantum channel Ψ is known as the recovery channel. A remarkable result of Knill and Laflamme[17] asserted that $\mathcal{C}$ is a k-dimensional QECC if and only if there exist $\gamma_{ij} \in \mathbb{C}$, $1 \leq i, j \leq r$, such that

$$P_{\mathcal{C}}(E_i^{\dagger} E_j) P_{\mathcal{C}} = \gamma_{ij} P_{\mathcal{C}} \quad \text{for all } 1 \leq i, j \leq r.$$

In Ref. 25, it was observed that this scheme can be improved in which no syndrome measurements, no syndrome readout ancillas and no projection operators were required, so that this scheme is relatively easy to implement. Under this modified scheme, for any quantum channel Φ of the form (1), there is a unitary recovery operation R for which the output state is a tensor product of the data qubit state and an encoding ancilla state. Suppose k is a multiple of $\dim \mathcal{H} = n$ and Φ has a k-dimensional quantum error correcting code $\mathcal{C}$. Then under certain orthonormal basis defined by $\mathcal{C}$, $\mathcal{H}$ can be decomposited into as $\mathcal{H}_A \otimes \mathcal{H}_B$ with dimension of $\mathcal{H}_A$ and $\mathcal{H}_B$ equal to n/k and k, respectively, so that there exist a unitary R and a positive semidefinite operator ξ such that

$$R^{\dagger} \Phi(|0\rangle\langle 0| \otimes \rho) R = \xi \otimes \rho \quad \text{for all} \quad \rho \in \mathcal{B}(\mathcal{H}_B).$$

As a result, a decoding scheme can be realized by a unitary operation followed by a partial trace operation.

Notice that by the classical result of Stinespring,[39] if the quantum states are represented by density operators acting on a Hilbert space $\mathcal{H}$, then for every quantum channel (trace preserving completely positive linear map) Φ, there is another Hilbert space $\mathcal{K}$ of much higher dimension than of $\mathcal{H}$, such that Φ can be dilated to a $*$-representation of $\mathcal{B}(\mathcal{K})$. Also it is worth noting that there are other automated QEC schemes which need ancillas for error detection, see for instance Ref. 30. An advantage of the scheme proposed in Ref. 25 is that there is no need to do the dilation as well as to introduce ancilla qubits for error detection, and only a unitary similarity transform is required. In some examples (see Refs. 23,25), one may use a permutation similarity transform, and a simple circuit diagram to implement the unitary similarity transform. Also the authors in Ref. 40 considered quantum channels where at most one single qubit is influenced by the Pauli matrix σ_x. More examples will be demonstrated in the next section.

2.2. Decoherence free subspace and noiseless subsystem

Decoherence free subspace[29,46–48] and Noiseless Subsystem[10,15,20,42] are another two standard methods to correct quantum error. A subspace $\mathcal{V}$ of $\mathcal{H}$ is said to be a decoherence free subspace (DFS) for a quantum channel Φ on $\mathcal{B}(\mathcal{H})$ if

$$\Phi(\rho) = \rho \quad \text{for all} \quad \rho \in \mathcal{B}(\mathcal{H}) \text{ with } \rho = P_{\mathcal{V}} \rho P_{\mathcal{V}},$$

where $P_{\mathcal{V}}$ is the orthogonal projection of $\mathcal{H}$ onto $\mathcal{V}$. Notice that a decoherence free subspace is a QECC with identity map as the recovery channel. In general, any QECC for a quantum channel Φ is a decoherence free subspace of the channel $\Psi \circ \Phi$. Different from the quantum error correcting code, the method of decherence free subspace is a passive error correction scheme. Data is stored in a special subspace so that it will not be affected by noise. However, the disadvantage of this scheme is that such decoherence free subspace may not exist.

A subsystem $\mathcal{H}_B$ is said to be a noiseless subsystem (NS) for a quantum channel Φ on $\mathcal{B}(\mathcal{H})$ if there are a co-subsystem $\mathcal{H}_A$ and a subspace $\mathcal{K}$ so that $\mathcal{H}$ has a decomposition $\mathcal{H} = (\mathcal{H}_A \otimes \mathcal{H}_B) \oplus \mathcal{K}$ in which for any $\rho^A \in \mathcal{B}(\mathcal{H}_A)$ and $\rho^B \in \mathcal{B}(\mathcal{H}_B)$, there is $\sigma^A \in \mathcal{B}(\mathcal{H}_A)$ such that

$$\Phi\left(\rho^A \otimes \rho^B\right) = \sigma^A \otimes \rho^B.$$

Given a decomposition $\mathcal{H} = (\mathcal{H}_A \otimes \mathcal{H}_B) \oplus \mathcal{K}$ and fixed an orthonormal basis $\{|a_1\rangle, \ldots, |a_p\rangle\}$ for $\mathcal{H}_A$. Let $P_{ij} = |a_i\rangle\langle a_j| \otimes \mathbf{1}_B$ for all $1 \leq i, j \leq p$ and $P_{AB} = P_{11} + \cdots + P_{pp}$. Notice that P_{AB} is the orthogonal projection of $\mathcal{H}$ onto $\mathcal{H}_A \otimes \mathcal{H}_B$. Then $\mathcal{H}_B$ is a noiseless subsystem for Φ if and only if

$$E_s P_{AB} = P_{AB} E_s P_{AB} \quad \text{for all} \quad 1 \leq s \leq r,$$

and there are scalars $\lambda_{i,j,s} \in \mathbb{C}$ such that

$$P_{ii} E_s P_{jj} = \lambda_{i,j,s} P_{ij} \quad \text{for all} \quad 1 \leq i, j \leq p, \ 1 \leq s \leq r.$$

In fact, a decoherence free subspace is indeed a special case of noiseless system, i.e., when $\dim \mathcal{H}_A = 1$. Then a subspace $\mathcal{V}$ of $\mathcal{H}$ is a decoherence free subspace for Φ if and only if there are scalars $\lambda_s \in \mathbb{C}$ such that

$$E_s P_{\mathcal{V}} = \lambda_s P_{\mathcal{V}} \quad \text{for all} \quad 1 \leq s \leq r.$$

2.3. Operator quantum error correction

In the paper of Kribs et al.,[21] the authors introduced a more generalized approach to quantum error correction. They call this scheme the operator quantum error correction (OQEC). A subsystem $\mathcal{H}_B$ is said to be a correctable subsystem (CS) for a quantum channel Φ on $\mathcal{B}(\mathcal{H})$ if there are a quantum channel Ψ on $\mathcal{B}(\mathcal{H})$, a co-subsystem $\mathcal{H}_A$, and a subspace $\mathcal{K}$ so that $\mathcal{H}$ has a decomposition $\mathcal{H} = (\mathcal{H}_A \otimes \mathcal{H}_B) \oplus \mathcal{K}$ and for any $\rho^A \in \mathcal{B}(\mathcal{H}_A)$ and $\rho^B \in \mathcal{B}(\mathcal{H}_B)$, there is $\sigma^A \in \mathcal{B}(\mathcal{H}_A)$ such that

$$\Psi \circ \Phi(\rho^A \otimes \rho^B) = \sigma^A \otimes \rho^B.$$

Or equivalently, Φ satisfies

$$\mathrm{tr}_A\left(P_{AB} \circ \Psi \circ \Phi(\rho^A \otimes \rho^B)\right) = \rho^B \text{ for all } \rho^A \in B(\mathcal{H}^A) \text{ and } \rho^B \in B(\mathcal{H}^B),$$

where P_{AB} is the orthogonal projection of $\mathcal{H}$ onto $\mathcal{H}_A \otimes \mathcal{H}_B$ as defined in the last subsection and where tr_A stands for the partial trace over the encoding ancilla subsystem $\mathcal{H}_A$. Notice that a noiseless subsystem is a correctable subsystem with identity map as the recovery channel. Also a correctable subsystem will reduce to a QECC if $\mathcal{H}_A$ has dimension 1. So this approach can be regarded as a unified formalism for all the technique mentioned in the last three subsections. A necessary and sufficient condition for the existence of correctable system was also given in Ref. 21 as follows. A subsystem $\mathcal{H}_B$ is a correctable subsystem for Φ if and only if there are scalars $\lambda_{i,j,s,t} \in \mathbb{C}$ such that

$$P_{ii}E_s^\dagger E_t P_{jj} = \lambda_{i,j,s,t}\, P_{ij} \quad \text{for all} \quad 1 \leq i, j \leq p,\ 1 \leq s, t \leq r.$$

3. Collective noise

Many physical systems have been proposed to build a working quantum computer, for example, liquid/solid-state NMR, trapped ions, linear optics, etc. In most settings, the quantum systems are microscopic in size, typically on the order of a few microns. On the other hand, the environmental noise, such as electromagnetic wave, has the wavelength on the order of a few centimeters or more. Therefore, one may assume that all the qubits in the register suffer from the same error operator. In this case, such error is called *collective error / noise*. Mathematically, the error operators of quantum channel with collective noise can be expressed as multiples of operators of the form $W^{\otimes n}$ with unitary W acting on a single qubit.

The quantum error correction for collective noise was studied by many researchers, see for example Refs. 27,28,41,42. Also quantum channel with corrective error of the form $\{\sigma_x^{\otimes n}, \sigma_y^{\otimes n}, \sigma_z^{\otimes n}\}$ were studied in Ref. 23 using the unitary recovery operation scheme, where $\sigma_x, \sigma_y, \sigma_z$ are the Pauli matrices. That is, all qubits constituting the codeword are subject to the

22

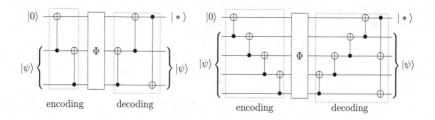

Fig. 1. Encoding and decoding circuits for n-qubit quantum channel with error operators $\left\{ \mathbf{1}^{\otimes n}, \sigma_x^{\otimes n}, \sigma_y^{\otimes n}, \sigma_z^{\otimes n} \right\}$ for $n = 3$ and 5. These circuits encode and recover an arbitrary $(n-1)$-qubit state $|\psi\rangle$ with a single ancilla qubit state $|0\rangle$ initially.

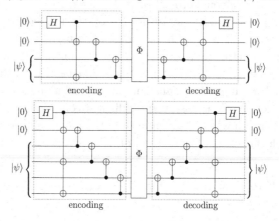

Fig. 2. Encoding and decoding circuits for n-qubit quantum channel with error operators $\left\{ \mathbf{1}^{\otimes n}, \sigma_x^{\otimes n}, \sigma_y^{\otimes n}, \sigma_z^{\otimes n} \right\}$ for $n = 4$ and 6. These circuits encode and recover an arbitrary $(n-2)$-qubit state $|\psi\rangle$ with a 2-qubit ancilla state $|00\rangle$ initially. Here, H is the Hadamard operator acting on a single qubit.

same Pauli operator in the channel. The authors showed that (i) n-qubit codeword can encodes $(n-1)$ data qubits when n is odd while (ii) n-qubit codeword implements a noiseless subsystem encoding $(n-2)$ data qubits when n is even. Also encoding and decoding circuits can been constructed systematically, see Figures 1 and 2 for n-qubit circuits for $n = 3, 4, 5, 6$. Quantum circuits with more qubits can be constructed in a similar strategy.

We now turn to the general case. Suppose the finite dimensional C^*-algebra $\mathcal{A}$ generated by the error operators $W^{\otimes n}$ in $\mathrm{SU}(2)^{\otimes n}$ admits the unique decomposition into irreducible representations up to unitary equivalence (similarity) as $\bigoplus_j (I_{r_j} \otimes M_{n_j})$ with $\sum_j r_j n_j = 2^n$, where n_j is

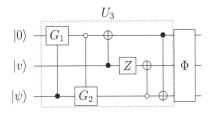

Fig. 3. Encoding circuit U_3 of 3-qubit NS, which encodes a qubit state $|\psi\rangle$. The ancilla state $|v\rangle$ can be arbitrary.

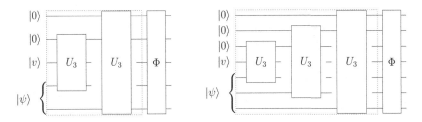

Fig. 4. (Left) Encoding circuit of 5-qubit NS, which encodes a two qubit state $|\psi\rangle$. (Right) Encoding circuit of 7-qubit NS, which encodes a three qubit state $|\psi\rangle$. Here U_3 is the circuit constructed in Figure 3.

the dimension of the irreducible representation while r_j its multiplicity. Then every error operator of the channel has the form $\bigoplus_j (I_{r_j} \otimes B_j)$ with $B_j \in M_{n_j}$. Now regard $M_{2^n} \cong (I_{r_j} \otimes M_{n_j}) \oplus M_q$ with $q = 2^n - r_j n_j$. According to this decomposition, when one applies the channel to a quantum state $\rho = (\tilde{\rho} \otimes \sigma) \oplus 0_q$ with $\tilde{\rho} \in M_{r_j}$ and $\sigma \in M_{n_j}$, $\Phi(\rho) = (\tilde{\rho} \otimes \sigma_E) \oplus 0_q$ because of the special decomposition of the error operators. Thus, the state $\tilde{\rho} \in M_{r_j}$ encoded as above will not be affected by the errors (noise) and can be easily recovered. This gives rise to a NS. The situation is simpler if $n_j = 1$, that is, we use the one-dimensional irreducible representations of $\mathcal{A}$, so that $\Phi(\tilde{\rho} \oplus 0_q) = \tilde{\rho} \oplus 0_q$. In such a case, we get a NFS. By the above discussion, construction of decoherence free subspace employs one-dimensional irreducible representations of the algebra $\mathcal{A}$ generated by $SU(2)^{\otimes n}$ for encoding while the construction of noiseless subsystem encodes logical qubits by makes use of the multiplicity of some irreducible representations.

In Ref. 24, the authors suggested the implementation of these ideas in terms of the quantum circuits. They have considered NFS with $n = 4$, which implements a single logical qubit and NS with $n = 3$ and 5, which encodes a single logical qubit and two logical qubits, respectively. Furthermore, simple encoding and decoding quantum circuits of NS for $n = 3$ and NFS for $n = 4$ have been constructed, see Figure 3 for the case $n = 3$. Here $G_1 = \frac{1}{\sqrt{3}} \begin{bmatrix} 1 & \sqrt{2} \\ -\sqrt{2} & 1 \end{bmatrix}$ and $G_2 = \frac{1}{\sqrt{2}} \begin{bmatrix} 1 & 1 \\ -1 & 1 \end{bmatrix}$. Viola *et al*[41] worked out the circuit implementation of $n = 3$ NS and demonstrated its validity by using an ion trap quantum computer. No further works have been conducted for $n \geq 4$ to date. Notice that the strategy in Ref.24 is to use the encoding/decoding circuit for $n = 3$ recursively in the implementation for all odd n so that $(n - 1)/2$-qubit state can be encoded in the circuit, see Figure 4 for the cases $n = 5$ and 7. Quantum circuits with more qubits can be constructed recursively. This is inspired by the recursive relations in the subspaces employed in NS and DFS. Recently, this scheme is extended to the study of quantum error correction for qudit.[14,26] It should be noted that the circuit for $n = 3$ is simpler than that obtained by Yang and Gea-Banacloche[42] and by Viola *et al*,[41] regarding the number of gates.

Acknowledgment

The authors would like to thank the organizers of the Summer Workshop on "Physics, Mathematics, and All That Quantum Jazz" for their excellent organization work and warm hospitality. The research of Li was supported by a USA NSF grant, a HK RGC grant, and the 2011 Shanxi 100 Talent Program. He is an honorary professor of University of Hong Kong, Taiyuan University of Technology, and Shanghai University. The research of Poon was supported by a USA NSF grant and a HK RGC grant. The research of Sze was supported by a HK RGC grant PolyU 502512.

References

1. P. Aliferis and A.W. Cross, Subsystem fault tolerance with the Bacon-Shor code, Phys. Rev. Lett. **98**, 220502 (2007).
2. T. Aoki, G. Takahashi, T. Kajiya, J. Yoshikawa, S.L. Braunstein, P. van Loock, and A. Furusawa, Quantum error correction beyond qubits, Nature Physics **5**, 541-546 (2009).
3. C.H. Bennett, D.P. DiVincenzo, J.A. Smolin, and W.K. Wootters, Mixed state entanglement and quantum error correction, Phys. Rev. A **54**, 3824-3851 (1996).
4. S.L. Braunstein, Quantum error correction of dephasing in 3 qubits, eprint arXiv:quant-ph/9603024.
5. A.R. Calderbank, E.M. Rains, P.W. Shor, and N.J.A. Sloane, Quantum error correction and orthogonal geometry, Phys. Rev. Lett. **78**, 405-408 (1997).

6. A.R. Calderbank, E.M. Rains, P.W. Shor, and N.J.A. Sloane, Quantum error correction via codes over $GF(4)$, IEEE Trans. Inf. Th. **44**, 1369-1387 (1998).

7. A.R. Calderbank and P.W. Shor, Good quantum error-correcting codes exist, Phys. Rev. A **54**, 1098-1105 (1996).

8. M.D. Choi, Completely positive linear maps on complex matrices, Linear Algebra Appl. **10**, 285-290 (1975).

9. D.G. Cory, M.D. Price, W. Maas, E. Knill, R. Laflamme, W.H. Zurek, T.F. Havel, and S.S. Somaroo, Experimental quantum error correction, Phys. Rev. Lett. **81**, 2152-2155 (1998).

10. S. De Filippo, Quantum computation using decoherence-free states of the physical operator algebra, Phys. Rev. A **62**, 052307 (2000).

11. L. DiCarlo, J.M. Chow, J.M. Gambetta, Lev S. Bishop, B.R. Johnson, D.I. Schuster, J. Majer, A. Blais, L. Frunzio, S.M. Girvin, and R.J. Schoelkopf, Demonstration of two-qubit algorithms with a superconducting quantum processor, Nature **460**, 240-244 (2009).

12. F. Gaitan, Quantum Error Correction and Fault Tolerant Quantum Computing, CRC Press, New York, 2008.

13. D. Gottesman, Class of quantum error-correcting codes saturating the quantum Hamming bound, Phys. Rev. A **54**, 1862-1868 (1996).

14. U. Güngördü, C.K. Li, M. Nakahara, Y.T. Poon, and N.S. Sze, Recursive encoding and decoding of the noiseless subsystem for qudits, eprint arXiv:quant-ph/1310.4401.

15. J. Kempe, D. Bacon, D.A. Lidar, and K.B. Whaley, Theory of decoherence-free fault-tolerant universal quantum computation, Phys. Rev. A **63**, 042307 (2001).

16. A.Y. Kitaev, A.H. Shen, and M.N. Vyalyi, Classical and Quantum Computation, Graduate Series in Mathematics no. 47, AMS, 2002.

17. E. Knill and R. Laflamme, Theory of quantum error-correcting codes, Phys. Rev. A **55**, 900-911 (1997).

18. Y. Kondo, C. Bagnasco, M. Nakahara, Quantum Error Correction with Uniformly Mixed State Ancillae, Phys. Rev. A **88**, 022314 (2013).

19. E. Knill, R. Laflamme, A. Ashikhmin, H. Barnum, L. Viola, and W.H. Zurek, Introduction to quantum error correction, eprint: arXiv:quant-ph/0207170.

20. E. Knill, R. Laflamme, and L. Viola, Theory of quantum error correction for general noise, Phys. Rev. Lett. **84**, 2525-2528 (2000).

21. D.W. Kribs, R. Laflamme, D. Poulin, and M. Lesosky, Operator quantum error correction, Quant. Inf. & Comp. **6**, 383-399 (2006).

22. R. Lang and P.W. Shor, Nonadditive quantum error correcting codes adapted to the amplitude damping channel, eprint: arXiv:quant-ph/0712.2586.

23. C.K. Li, M. Nakahara, Y.T. Poon, N.S. Sze, and H. Tomita, Efficient quantum error correction for fully correlated noise, Phys. Lett. A **375**, 3255-3258 (2011).

24. C.K. Li, M. Nakahara, Y.T. Poon, N.S. Sze, and H. Tomita, Recursive encoding and decoding of noiseless subsystem and decoherence free subspace, Phys. Rev. A **84**, 044301 (2011).

25. C.K. Li, M. Nakahara, Y.T. Poon, N.S. Sze, and H. Tomita, Recovery in

quantum error correction for general noise without measurement, Quant. Inf. & Comp. **12**, 149-158, (2012).

26. C.K. Li, M. Nakahara, Y.T. Poon, and N.S. Sze, Recursive encoding and decoding of the noiseless subsystem for qudits, eprint arXiv:quant-ph/1306.0981.

27. X.H. Li, F.G. Deng, and H.Y. Zhou, Faithful qubit transmission against collective noise without ancillary qubits, Appl. Phys. Lett. **91**, 144101 (2007).

28. D.A. Lidar, D. Bacon, and K.B. Whaley, Concatenating decoherence-free subspaces with quantum error correcting codes, Phys. Rev. Lett. **82**, 4556-4559 (1999).

29. D.A. Lidar, I.L. Chuang, and K.B. Whaley, Decoherence-free subspaces for quantum computation, Phys. Rev. Lett. **81**, 2594-2597 (1998).

30. N.D. Mermin, Quantum Computer Science: An Introduction, Cambridge University Press, 2007.

31. M. Nakahara and T. Ohmi, Quantum Computing, From Linear Algebra to Physical Realization, CRC Press, New York, 2008.

32. M.A. Neilsen and I.L. Chuang, Quantum Computation and Quantum Information, Cambridge University Press, Cambridge, 2000.

33. J. Preskill, Reliable quantum computers, Proc. R. Soc. A **454**, 385-410 (1998).

34. M.D. Reed, L. DiCarlo, S.E. Nigg, L. Sun, L. Frunzio, S.M. Girvin, and R.J. Schoelkopf, Realization of three-qubit quantum error correction with superconducting circuits, Nature **482**, 382-385 (2012).

35. P.W. Shor, Scheme for reducing decoherence in quantum computer memory, Phys. Rev. A **52**, 2493-2496 (1995).

36. J.A. Smolin, G. Smith, and S. Wehner, Simple family of nonadditive quantum codes, Phys. Rev. Lett. **99**, 130505 (2007).

37. A.M. Steane, Error correcting codes in quantum theory, Phys. Rev. Lett. **77**, 793-797 (1996).

38. A.M. Steane, Decoherence and Its Implications in Quantum Computation and Information Transfer, IOS Press, Amsterdam, 2001.

39. W.F. Stinespring, Positive functions on C^*-algebras, Pro. Amer. Math. Soc. **6**, 211-216 (1955).

40. H. Tomita and M. Nakahara, Unitary quantum error correction without error detection, eprint arXiv:quant-ph/1101.0413.

41. L. Viola, E.M. Fortunato, M.A. Pravia, E. Knill, R. Laflamme, and D.G. Cory, Experimental realization of noiseless subsystems for quantum information processing, Science **293**, 2059-2063 (2001).

42. C.P. Yang and J. Gea-Banacloche, Three-qubit quantum error-correction scheme for collective decoherence, Phys. Rev. A **63**, 022311 (2001).

43. W.T. Yen and L.Y. Hsu, Optimal nonadditive quantum error-detecting code, eprint arXiv:quant-ph/0901.1353.

44. S. Yu, Q. Chen, and C.H. Oh, Two infinite families of nonadditive quantum error-correcting codes, eprint arXiv:quant-ph/0901.1935.

45. S. Yu, Q. Chen, C.H. Lai, and C.H. Oh, Nonadditive quantum error-correcting code, Phys. Rev. Lett. **101**, 090501 (2008).

46. P. Zanardi and M. Rasetti, Noiseless quantum codes, Phys. Rev. Lett. **79**, 3306-3309 (1997).
47. P. Zanardi and M. Rasetti, Error avoiding quantum codes, Mod. Phys. Lett. B **11**, 1085 (1997).
48. P. Zanardi, Dissipation and decoherence in a quantum register, Phys. Rev. A **57**, 3276-3284 (1998).

IMPLEMENTING MEASUREMENT OPERATORS IN LINEAR OPTICAL AND SOLID-STATE QUBITS

YUKIHIRO OTA[1], SAHEL ASHHAB[1] and FRANCO NORI[1,2]

[1] *CEMS, RIKEN,*
2-1 Hirosawa, Saitama, 351–0198, Japan
[2] *Physics Department, University of Michigan,*
Ann Arbor, MI, 48109-1040, USA

We show a systematic construction for implementing general measurements on a single qubit, including both strong (or projection) and weak measurements. We mainly focus on linear optical qubits. The present approach is composed of simple and feasible elements, i.e., beam splitters, wave plates, and polarizing beam splitters. We show how the parameters characterizing the measurement operators are controlled by the linear optical elements. We also propose a method for the implementation of general measurements in solid-state qubits.

Keywords: Quantum measurements; Linear optics; Superconducting qubits

1. Introduction

A quantum measurement is formulated as an operation to extract information from a quantum system (see, e.g., Refs. 1–4). This operation causes a disturbance to the system. The amount of information extracted from the system can be related to the measurement back-action.[5–8] This intriguing feature of quantum measurements leads to measurement-based methods in quantum engineering. Koashi and Ueda[9] proposed a reversible quantum measurement. Nakazato *et al.*[10] developed a method for producing a pure state from a arbitrary mixed state, with a sequence of strong measurements. Stockton *et al.*[11] showed a deterministic method for preparing an entangled state, with continuous measurement. Lupaşcu *et al.*[12] implemented a quantum non-demolition measurement in a superconducting flux qubit. Korotkov and Keane[13] invented a way to suppress decoherence, with a reversal measurement. Two of the authors (SA and FN)[14] proposed a control-free quantum control, with partial measurements. Also, the authors[15] proposed an entanglement amplification protocol for a single copy of a partially-entangled pure state. Inoue *et al.*[16] implemented a quantum-noise suppression through measurement-based feedback, in ultra-cold atom gases.

A large number of experimental studies on quantum measurements have been performed. Huttner *et al.* [17] discriminated with two nonorthogonal states, using linear optical systems. Gillett *et al.* [18] demonstrated quantum feedback control for a photonic polarization qubit. A polarizing beam splitter with a tunable reflection coefficient was used on the above-mentioned two experiments. Kwiat *et al.* [19] implemented an entanglement concentration protocol with a partial-collapse measurement in a photonic qubit. Kim *et al.* [20] demonstrated a reversal operation of a weak measurement for a photonic qubit. The key idea in the two experiments is to use Brewster-angle glass plates. Katz *et al.* [21] performed a conditional recovery of a quantum state with a partial-collapse measurement in Josephson phase qubits. Iinuma *et al.* [22] studied the observation of a weak value. Kocsis *et al.* [23] examined a photon's "trajectories" through a double-slit interferometer with weak measurements of the photon momentum.

A general (or weak) measurement is associated with a positive-operator valued measure (POVM) .[2] A POVM on a measured quantum system can be expressed as a projection-valued measure in an extended system including the target system and an ancillary (or probe) system, as seen in, e.g., Ref. 4. Therefore, we may construct arbitrary measurements from von Neumann measurements on the extended system. However, this statement does not give a specific and simple recipe to design measurement operators. Hence, a systematic approach to realize this general idea in specific physical systems is highly desirable.

In this paper, we show a method [24] to implement general measurements on a single qubit, including both von Neumann and weak measurements. We mainly focus on a linear optical qubit. Depending on the path degree of freedom in an interferometer, a polarization state is transformed by a non-projective positive operator. We show a systematic prescription for designing various measurement operators. The present approach is composed of simple and basic linear optical elements, i.e., beam splitters, wave plates, and polarizing beam splitters. We show how the parameters characterizing the measurement operators (e.g., the measurement strength) can be tuned. We also show a method for implementing general measurements on a solid-state qubit, especially a superconducting qubit.

2. Setting

Let us consider a linear optical system with N input and N output modes. The following arguments are also applicable to other physical systems such as neutron interferometers [25] and on-chip photonics.[26] The annihilation (creation) operator in the nth input mode is defined as $\hat{a}_{\text{in},n}$ ($\hat{a}^\dagger_{\text{in},n}$) with $n = 1, 2, \ldots, N$. The vacuum $|0\rangle$ is defined as $\hat{a}_{\text{in},n}|0\rangle = 0$. The bosonic canonical commutation relations are

$$[\hat{a}_{\text{in},m}, \hat{a}^\dagger_{\text{in},n}] = \delta_{nm}, \quad [\hat{a}_{\text{in},m}, \hat{a}_{\text{in},n}] = [\hat{a}^\dagger_{\text{in},m}, \hat{a}^\dagger_{\text{in},n}] = 0. \tag{1}$$

Similarly, we define the annihilation (creation) operator in the nth output mode as $\hat{a}_{\text{out},n}$ ($\hat{a}^\dagger_{\text{out},n}$). A single photon with arbitrary polarization enters the system from the first input mode. The polarization is described by a vector $|\psi\rangle \in \mathbb{C}^2$, with $\langle\psi|\psi\rangle = 1$. We assume that the input state is a pure state $|\Psi\rangle = |\psi\rangle \otimes \hat{a}^\dagger_{\text{in},1}|0\rangle$. In this setting, the photon polarization is measured, using the path degree of freedom in the interferometer as an ancillary qubit.

We calculate the photon state at each output mode. The field operators at the output modes are related to the input modes via a linear canonical transformation [27–30] because the system is composed of linear optical elements (e.g., beam splitters). The polarization of the photon at the nth output mode is described by

$$|\phi_n\rangle = \langle 0| \hat{a}_{\text{out},n}|\Psi\rangle. \tag{2}$$

Thus, we find that in terms of the output mode operators the photon state is written as

$$|\Psi^{(\text{out})}\rangle = \sum_{n=1}^{N} |\phi_n\rangle \otimes \hat{a}^\dagger_{\text{out},n}|0\rangle = \sum_{n=1}^{N} \hat{X}_n|\psi\rangle \otimes \hat{a}^\dagger_{\text{out},n}|0\rangle, \tag{3}$$

with a linear operator $\hat{X}_n$ on $\mathbb{C}^2$ such that $|\phi_n\rangle = \hat{X}_n|\psi\rangle$. Since there is no photon loss, we have $\langle\Psi^{\text{out}}|\Psi^{\text{out}}\rangle = \langle\Psi|\Psi\rangle$. The normalization condition $\langle\Psi|\Psi\rangle = 1$ implies that $\sum_{n=1}^{N} \hat{X}^\dagger_n\hat{X}_n = \hat{I}_2$, where the identity operator on $\mathbb{C}^2$ is $\hat{I}_2$. Therefore, we find that the initial state $|\psi\rangle$ is transformed by a linear operator which is related with an element of a POVM.

It is convenient for designing various kinds of POVM to add another linear optical system to some of the output modes in the interferometer before the photon detection. For example, let us set wave plates in each output mode. The wave plates in the nth mode describe a unitary gate V_n given as the unitary part of $\hat{X}_n$. Using the right-polar decomposition,[31] we find that $\hat{X}_n = \hat{V}^\dagger_n\hat{M}_n$, where $\hat{M}_n = (\hat{X}^\dagger_n\hat{X}_n)^{1/2}$ is a positive operator. The role of the wave plates is to remove the effect of the unitary evolution $\hat{V}_n^\dagger$ from $\hat{X}_n$. Using the application of such "path-dependent" unitary operators, we find that

$$|\Psi^{(\text{out})}\rangle \rightarrow \sum_{n=1}^{N} |\psi_n\rangle \otimes \hat{a}_{\text{out},n}^\dagger |0\rangle, \tag{4}$$

where $|\psi_n\rangle = \hat{V}_n|\phi_n\rangle = \hat{M}_n|\psi\rangle$. The measurement operators $\hat{M}_n$ satisfy $\sum_{n=1}^{N} \hat{M}_n^\dagger \hat{M}_n = \hat{I}_2$. Thus, we have the measurements corresponding to the POVM $\{\hat{E}_n\}_{n=1}^{N}$ with $\hat{E}_n = \hat{M}_n^\dagger \hat{M}_n$. The positive operator $\hat{M}_n$ is the essential part (i.e., the back-action associated with the measurement) of non-unitarity of $\hat{X}_n$ and called minimally-disturbing measurement, following Ref. 8. Throughout this paper, we will focus on such minimally-disturbing measurements.

Let us characterize the measurement operators $\hat{M}_n$. We can expand $\hat{M}_n$ as $\hat{M}_n = \sum_{i=1}^{2} m_{n,i}|m_{n,i}\rangle\langle m_{n,i}|$, with the eigenvalues $m_{n,i} (\geq 0)$ and the associated eigenvectors $|m_{n,i}\rangle$. The eigenvalues are related to the measurement strength, while the eigenvectors are regarded as the measurement direction. Let us consider the case when $m_{n,1} = 1$ and $m_{n,2} = 0$, for example. We find that $\hat{M}_n$ is a projection operator (i.e., a sharp measurement) in the direction of $\vec{m}_{n,1}$, where $\vec{m}_{n,1}$ is the Bloch vector corresponding to $|m_{n,1}\rangle\langle m_{n,1}|$. The indistinguishability between the elements of a POVM is characterized by the Hilbert-Schmidt inner product $\text{Tr}\,\hat{E}_n^\dagger \hat{E}_m$ for $n \neq m$. The elements of a projection-valued measure are distinguishable (i.e., the inner products are zero), for example. Now, we pose the question: How are these measurement features controlled by typical linear optical devices? We will answer this question in the next section.

3. Design of general measurements on linear optical qubits

3.1. Symmetric arbitrary-strength two-outcome measurements

Let us first define a special class of general measurements, which is the target operation in this subsection. We consider a symmetric two-outcome POVM in $\mathbb{C}^2$ composed of convex combinations of two orthogonal projection operators

$$\hat{M}_1 = \sqrt{\frac{1+\varepsilon}{2}}|m_+\rangle\langle m_+| + \sqrt{\frac{1-\varepsilon}{2}}|m_-\rangle\langle m_-|, \tag{5}$$

$$\hat{M}_2 = \sqrt{\frac{1-\varepsilon}{2}}|m_+\rangle\langle m_+| + \sqrt{\frac{1+\varepsilon}{2}}|m_-\rangle\langle m_-|, \tag{6}$$

with $0 \leq \varepsilon \leq 1$ and $\langle m_+|m_-\rangle = 0$. Adjusting a real parameter ε allows arbitrary measurement strength. We find that if the coefficient for $|m_+\rangle\langle m_+|$ increases, then the one for $|m_-\rangle\langle m_-|$ decreases, and vice versa. Thus, these measurement operators are symmetric or balanced with respect to the parameter ε. We call the measurements described by these linear operators symmetric arbitrary-strength two-outcome measurments (SASTOM) on $\mathbb{C}^2$. The disturbance induced by a SASTOM is considered to be minimum, as shown in Ref. 6. Another important property of a SASTOM is $\mathrm{Tr}\, \hat{M}_1^\dagger \hat{M}_1 = \mathrm{Tr}\, \hat{M}_2^\dagger \hat{M}_2 = 1$. An application of a SASTOM to quantum protocols is shown in, e.g., Ref. 15.

Now, we propose a systematic way to construct this measurement. Let us consider an interferometer with two input and two output modes. A single photon enters a polarizing beam splitter [32] from the first input mode. Subsequently, its polarization at each arm is transformed by wave plates. Then, the two modes are recombined in a beam splitter. Let us apply the basic idea in Sec. 2 to this interferometer. We need three kinds of field operators for describing the path degrees of freedom, i.e., input, intermediate, and output field operators. We have an initial pure state

$$|\Psi^{(\mathrm{in})}\rangle = |\psi\rangle \otimes \hat{a}_{\mathrm{in},1}^\dagger |0\rangle, \quad |\psi\rangle = c_{\mathrm{H}}|\mathrm{H}\rangle + c_{\mathrm{V}}|\mathrm{V}\rangle,$$

with the horizontal polarization state $|\mathrm{H}\rangle$ and the vertical polarization state $|\mathrm{V}\rangle$. The field operators $\hat{a}_{\mathrm{int},n}$ and $\hat{a}_{\mathrm{int},n}^\dagger$ at the output modes of the polarizing beam splitter are related to $\hat{a}_{\mathrm{in},n}$ and $\hat{a}_{\mathrm{in},n}^\dagger$ via $\hat{a}_{\mathrm{int},\mathrm{H},1} = \hat{a}_{\mathrm{in},\mathrm{H},1}$, $\hat{a}_{\mathrm{int},\mathrm{H},2} = \hat{a}_{\mathrm{in},\mathrm{H},2}$, $\hat{a}_{\mathrm{int},\mathrm{V},1} = \hat{a}_{\mathrm{in},\mathrm{V},2}$, and $\hat{a}_{\mathrm{int},\mathrm{V},2} = \hat{a}_{\mathrm{in},\mathrm{V},1}$,[30] where $\hat{a}_{\mathrm{in},\mathrm{H},1} = |\mathrm{H}\rangle\langle\mathrm{H}| \otimes \hat{a}_{\mathrm{in},1}$, etc. After the single photon passes through the polarizing beam splitter, we find that the photon state is written as $|\Psi^{(\mathrm{int})}\rangle = |u_1\rangle \otimes \hat{a}_{\mathrm{int},1}^\dagger |0\rangle + |u_2\rangle \otimes \hat{a}_{\mathrm{int},2}^\dagger |0\rangle$ with $|u_1\rangle = c_{\mathrm{H}}|\mathrm{H}\rangle$ and $|u_2\rangle = c_{\mathrm{V}}|\mathrm{V}\rangle$. We remark that the polarizing beam splitter produces entanglement between the polarization and the path degrees of freedom in the interferometer. Next, we perform unitary operations for the polarization on each of these intermediate modes with the wave plates. The use of half- and quarter-wave plates allows the construction of arbitrary elements of SU(2).[33,34] Hence, using the path-dependent wave plates, we find that

$$|\Psi^{(\mathrm{int})}\rangle \rightarrow \sum_{n=1}^{2} \hat{U}_n |u_n\rangle \otimes \hat{a}_{\mathrm{int},n}^\dagger |0\rangle,$$

with $\hat{U}_n \in \mathrm{SU}(2)$. The canonical transformation associated with the beam splitter [29,30] is $\hat{a}_{\mathrm{out},1} = r\hat{a}_{\mathrm{int},1} + t\hat{a}_{\mathrm{int},2}$ and $\hat{a}_{\mathrm{out},2} = t\hat{a}_{\mathrm{int},1} - r\hat{a}_{\mathrm{int},2}$, where $r^2 + t^2 = 1$ and $0 \leq r, t \leq 1$. After the single photon passes through the beam splitter, we find that the photon state is expressed by

$$|\Psi^{(\text{out})}\rangle = \sum_{n=1}^{2} |\phi_n\rangle \otimes \hat{a}_{\text{out},n}^{\dagger}|0\rangle,$$

where $|\phi_1\rangle = r\hat{U}_1|u_1\rangle + t\hat{U}_2|u_2\rangle$ and $|\phi_2\rangle = t\hat{U}_1|u_1\rangle - r\hat{U}_2|u_2\rangle$. We find that the linear operators satisfying $|\phi_n\rangle = \hat{X}_n|\psi\rangle$ are $\hat{X}_1 = r\hat{U}_1|\text{H}\rangle\langle\text{H}| + t\hat{U}_2|\text{V}\rangle\langle\text{V}|$ and $\hat{X}_2 = t\hat{U}_1|\text{H}\rangle\langle\text{H}| - r\hat{U}_2|\text{V}\rangle\langle\text{V}|$. The positive operators associated with $\hat{X}_1$ and $\hat{X}_2$ are given, respectively, by Eqs. (5) and (6) with

$$\varepsilon = \sqrt{1 - 4r^2t^2(1 - |w|^2)}, \quad w = \langle\text{H}|\hat{U}_1^{\dagger}\hat{U}_2|\text{V}\rangle. \tag{7}$$

The definition of w implies that the choice of the unitary gates $\hat{U}_1$ and $\hat{U}_2$ is not unique for constructing the measurement operators $\hat{M}_n$. For example, one can set $\hat{U}_1$ as the identity operator and still obtain any desired value of w by adjusting $\hat{U}_2$. Iinuma et al. [22] set $\hat{U}_1 = \hat{U}_2^{\dagger}$, which results in a real value for w and thus imposes a constraint on the measurement operators that can be constructed. The range of $|w|$ is $0 \leq |w| \leq 1$ since w is an off-diagonal element of a unitary operator. The basis vectors $|m_+\rangle$ and $|m_-\rangle$ for $w \neq 0$ are

$$|m_+\rangle = \cos\frac{\theta}{2}|\text{H}\rangle + e^{-i\phi/2}\sin\frac{\theta}{2}|\text{V}\rangle, \tag{8}$$

$$|m_-\rangle = -e^{i\phi/2}\sin\frac{\theta}{2}|\text{H}\rangle + \cos\frac{\theta}{2}|\text{V}\rangle, \tag{9}$$

where $\tan(\theta/2) = (t^2 - r^2 + \varepsilon)/2rt|w|$ and $e^{i\phi/2} = w/|w|$. When $w = 0$ and $r^2 \geq t^2$ $(r^2 < t^2)$, we have $|m_+\rangle = |\text{H}\rangle$ and $|m_-\rangle = |\text{V}\rangle$ $(|m_+\rangle = |\text{V}\rangle$ and $|m_-\rangle = -|\text{H}\rangle)$. The fact that $[\hat{X}_1^{\dagger}\hat{X}_1, \hat{X}_2^{\dagger}\hat{X}_2] = 0$ indicates that $\hat{M}_1$ and $\hat{M}_2$ are simultaneously diagonalizable. The expression for $\hat{V}_n$ such that $\hat{X}_n = \hat{V}_n^{\dagger}\hat{M}_n$ is calculated straightforwardly. After the applications of $\hat{V}_n$, we find that

$$|\Psi^{(\text{out})}\rangle \rightarrow \sum_{n=1}^{2} \hat{M}_n|\psi\rangle \otimes \hat{a}_{\text{out},n}^{\dagger}|0\rangle.$$

The tunable parameter ε is the measurement strength and completely determines the indistinguishability between the POVM elements,

$$\text{Tr}\,\hat{E}_1^{\dagger}\hat{E}_2 = \frac{1}{2}(1 - \epsilon^2). \tag{10}$$

An application of the SASTOM is a way of entanglement concentration.[15] Applications of the local SASTOM to a partially-entangled pure state lead to a probablistic amplification of the entanglement. The key idea is to tune the measurement strength at each measurement step. Subsequently, we will briefly illustrate this method.

Let us consider a bipartite system composed of a two- and a d-level system ($d \geq 2$). The two-level system is called system A, while the d-level one is called system B. We apply some operations only to system A and never touch system B. Since system A is a single-qubit, the treatment of control processes is simplified. We have a pure state $|\psi\rangle$ in this composite system,

$$|\psi\rangle = \alpha |0\rangle |\phi_0\rangle + \beta |1\rangle |\phi_1\rangle, \quad |i\rangle \in \mathbb{C}^2, \quad |\phi_i\rangle \in \mathbb{C}^d, \tag{11}$$

with $\alpha^2 + \beta^2 = 1$, $\alpha \geq 0$, $\beta \geq 0$, $\langle i|j\rangle = \delta_{ij}$, and $\langle \phi_i|\phi_j\rangle = \delta_{ij}$ ($i, j = 0, 1$). Mathematically, we can always find this form of a vector with the Schmidt decomposition.[4] We show a method to distill entanglement from an initial partially-entangled pure state. The elementary part of the scheme is the SASTOM on system A. The resultant state becomes $\rho_{\text{out}} = p_1 |\psi_1\rangle\langle\psi_1| + p_2 |\psi_2\rangle\langle\psi_2|$ with $|\psi_\ell\rangle = (\hat{M}_\ell \otimes \hat{I}_d)|\psi\rangle/\sqrt{p_\ell}$ and $p_\ell = \langle\psi|(\hat{M}_\ell^\dagger \hat{M}_\ell \otimes \hat{I}_d)|\psi\rangle$ ($\ell = 1, 2$). The important point here is that $|\psi_1\rangle$ is a maximally-entangled state when $\alpha^2(1 + \varepsilon)/2 = \beta^2(1 - \varepsilon)/2$. Assuming that $\beta > \alpha$, we find that the parameter ε for obtaining a maximally-entangled state becomes

$$\varepsilon = \beta^2 - \alpha^2. \tag{12}$$

On the basis of the above arguments, we propose a probabilistic method to make a maximally-entangled state from $|\psi\rangle$ with the successive application of the local SASTOM's. We prepare the N weak measurement apparatus for the SASTOM. We stop the measurement process once if we obtain the outcome $\hat{M}_1^\dagger \hat{M}_1$. Otherwise, we move to the subsequent measurement. The input state in the nth measurement apparatus is

$$|\psi_n\rangle = \frac{1}{p_2(\varepsilon_{n-1})}[\hat{M}_2(\varepsilon_{n-1}) \otimes \hat{I}_d]|\psi_{n-1}\rangle,$$

$$p_2(\varepsilon_{n-1}) = \langle\psi_{n-1}|\hat{M}_2(\varepsilon_{n-1})^\dagger \hat{M}_2(\varepsilon_{n-1}) \otimes \hat{I}_d|\psi_{n-1}\rangle,$$

if $n \geq 2$ and $|\psi_1\rangle = |\psi\rangle$, with $\beta > \alpha$. In addition, we find that $\beta_n > \alpha_n$ if $\beta_{n-1} > \alpha_{n-1}$ and $\varepsilon_n > 0$. Then, when the measurement strength obeys the recurrence relation

$$\varepsilon_n = \frac{2\varepsilon_{n-1}}{1 + (\varepsilon_{n-1})^2} \quad (n \geq 2), \quad \varepsilon_1 = \beta^2 - \alpha^2, \tag{13}$$

the state corresponding to the outcome $\hat{M}_1^\dagger(\varepsilon_n)\hat{M}_1(\varepsilon_n) \otimes \hat{I}_d$ is the maximally-entangled state $(|0\rangle|\phi_0\rangle + |1\rangle|\phi_1\rangle)/\sqrt{2}$. The probability to obtain the outcome $\hat{M}_1^\dagger \hat{M}_1$ at the nth measurement apparatus is

$$P_n = \frac{1}{2}[1 - (\varepsilon_n)^2]. \tag{14}$$

When $\beta < \alpha$, we can obtain the same result replacing the role of $\hat{M}_1$ with $\hat{M}_2$. Since ε_n obeys the recurrence relation (13), the information required is only $\beta^2 - \alpha^2$.

3.2. General two-outcome measurements

Various quantum protocols with measurement operators not expressed by Eqs. (5) and (6) have been proposed (e.g., Refs. 13,14). We extend the approach developed in Sec. 3.1 for implementing such general measurements. Two generalization routes may exist. One involves increasing the number of parameters characterizing the two measurement operators $\hat{M}_1$ and $\hat{M}_2$. The other is to increase the number of outcomes. First, we examine the former. The eigenvalues of the measurement operators (5) and (6) are parametrized by ε. The measurement direction contains the two parameters θ and ϕ. Thus, the number of parameters in the measurement operators is equal to a density matrix on $\mathbb{C}^2$. This point is also confirmed by the fact that $\operatorname{Tr} \hat{M}_n^\dagger \hat{M}_n = 1$. Since the positive operator $\hat{M}_n^\dagger \hat{M}_n$ is a density matrix on $\mathbb{C}^2$, its square root $\hat{M}_n$ is characterized by three real parameters.

To implement general two-outcome measurements, we add another interferometer with reflection coefficient r' and transmission coefficient t' to the output arms in the interferometer for a SASTOM. Accordingly, we need four kinds of field operators for describing the path degrees of freedom. The calculations before the last beam splitter are the same as in Sec. 3.1. Namely, the first polarizing beam splitter creates entanglement between the polarization and the path degree of freedom. The subsequent wave plates transform photon's polarization through unitary operators depending on the path degree of freedom. The corresponding unitary operator $\hat{U}_n$ is an arbitrary element of SU(2), as seen in Sec. 3.1. Then, the two paths are recombined in the intermediate beam splitter with reflection coefficient r. At this point, the polarization state in the nth mode is described by $\hat{X}_n |\psi\rangle$. After the photon passes through this intermediate beam splitter, the unitary operators $\hat{V}_1$ and $\hat{V}_2$ are applied to the first and the second paths, respectively, in order to remove the unitary parts of $\hat{X}_1$ and $\hat{X}_2$. These unitary operators $\hat{V}_1$ and $\hat{V}_2$ are automatically determined when calculating the right-polar decompositions of $\hat{X}_1$ and $\hat{X}_2$. Thus, the resultant state in the nth mode becomes $\hat{M}_n |\psi\rangle$, as seen in Eqs. (5) and (6). As shown in

Sec. 3.1, in this step, we have three tunable parameters, i.e., r, the modulus of w, and the phase of w, where $w = \langle H|\hat{U}_1^\dagger \hat{U}_2|V\rangle$.

Let us now consider what happens in the additional part. We write the input (output) field operators in the last beam splitter as $\hat{a}'_{\text{int},n}$ and $\hat{a}'^\dagger_{\text{int},n}$ ($\hat{a}_{\text{out},n}$ and $\hat{a}^\dagger_{\text{out},n}$). The related linear canonical transformations is $\hat{a}_{\text{out},1} = r'\hat{a}'_{\text{int},1} + t'\hat{a}'_{\text{int},2}$ and $\hat{a}_{\text{out},2} = t'\hat{a}'_{\text{int},1} - r'\hat{a}'_{\text{int},2}$ with $(r')^2 + (t')^2 = 1$ and $0 \le r', t' \le 1$. After the photon passes through the last beam splitter, its state is written as $|\Psi^{(\text{out})}\rangle = \sum_{n=1}^{2} \hat{X}'_n |\psi\rangle \otimes \hat{a}^\dagger_{\text{out},n} |0\rangle$, where $\hat{X}'_1 = r'\hat{M}_1 + t'\hat{M}_2$ and $\hat{X}'_2 = t'\hat{M}_1 - r'\hat{M}_2$. The positive-operator parts of $\hat{X}'_1$ and $\hat{X}'_2$ are, respectively,

$$\hat{M}'_1 = \sqrt{p}|m_+\rangle\langle m_+| + \sqrt{q}|m_-\rangle\langle m_-|, \tag{15}$$
$$\hat{M}'_2 = \sqrt{1-p}|m_+\rangle\langle m_+| + \sqrt{1-q}|m_-\rangle\langle m_-|., \tag{16}$$

with

$$\sqrt{p} = r'\sqrt{\frac{1+\varepsilon}{2}} + t'\sqrt{\frac{1-\varepsilon}{2}}, \tag{17}$$

$$\sqrt{q} = r'\sqrt{\frac{1-\varepsilon}{2}} + t'\sqrt{\frac{1+\varepsilon}{2}}. \tag{18}$$

The measurement direction is the same as the SASTOM in the previous subsection. This means that the basis vectors $|m_+\rangle$ and $|m_-\rangle$ do not depend on the parameter r'. We remark that $\hat{M}'_1 = \hat{X}'_1$ since $\hat{X}'_1$ is a positive operator. The measurement operator $\hat{M}'_2$ is related with the linear operator $\hat{X}'_2$ via a unitary operator, $\hat{M}'_2 = \hat{S}\hat{X}'_2$, with $\hat{S}^\dagger \hat{S} = \hat{S}\hat{S}^\dagger = \hat{I}_2$. With some algebra one can show that when

$$\sqrt{1-\varepsilon^2} \le 2r't',$$

the unitary operator $\hat{S}$ is a phase-shift gate (i.e., $\hat{S} = |m_+\rangle\langle m_+| - |m_-\rangle\langle m_-|$). Otherwise, $\hat{S}$ is equal to the identity operator, up to an overall phase. The measurement operators $\hat{M}'_1$ and $\hat{M}'_2$ are characterized by two independent positive parameters p and q ($0 \le p, q \le 1$). In contrast to a SASTOM, the trace of $\hat{M}'^\dagger_n \hat{M}'_n$ is not fixed. We find that $\text{Tr}\,\hat{M}'^\dagger_1 \hat{M}'_1 = 1 + \Delta$ and $\text{Tr}\,\hat{M}'^\dagger_2 \hat{M}'_2 = 1 - \Delta$, with $\Delta = p + q - 1$. The indistinguishability between the elements of the corresponding POVM is

$$\text{Tr}\,\hat{E}'^\dagger_1 \hat{E}'_2 = p(1-p) + q(1-q), \tag{19}$$

where $\hat{E}'_n = \hat{M}'^\dagger_n \hat{M}'_n$. We stress that all of the features in the measurement operators are tunable via the basic linear optical elements. We also remark that the action of Eqs. (15) and (16) can be obtained by a polarizing beam splitter with tunable reflection coefficients, which has been used for implementing general two-outcome measurements in optical setups.[17,18]

3.3. *General multi-outcome measurements*

Next, we examine another generalization of the SASTOM on $\mathbb{C}^2$. The repeated application of general two-outcome measurements allows the construction of a POVM with multiple outcomes. Let us consider a system composed of $(N - 1)$ detectors, each of which performs a general two-outcome measurement. At the ℓth detector, the first output mode corresponds to an outcome, while the second output mode is regarded as an input mode for the subsequent device. Thus, we find that the entire system has N outcomes. This "branch structure" is one possible realization of a multi-outcome POVM.

Now, let us show how a general multi-outcome measurement is implemented. Let us write the measurement operators in the ℓth apparatus as $\hat{M}_1^{(\ell)}$ and $\hat{M}_2^{(\ell)}$. Their expressions are given in Eqs. (15) and (16). The measurement operator corresponding to the ℓth outcome is written as $\hat{K}_\ell$ ($\ell = 1, 2, \ldots, N$) and is constructed recursively using

$$\hat{K}_\ell = \hat{W}_\ell \hat{M}_1^{(\ell)} \hat{Y}_\ell \quad (2 \leq \ell \leq N - 1), \tag{20}$$

$$\hat{K}_1 = \hat{M}_1^{(1)}, \quad \hat{K}_N = \hat{W}_N \hat{Y}_N. \tag{21}$$

The linear operator $\hat{Y}_\ell$ is defined as $\hat{Y}_\ell = \hat{M}_2^{(\ell-1)} \hat{Y}_{\ell-1}$ ($\ell \geq 2$) with $\hat{Y}_1 = \hat{I}_2$. We remark that $\hat{Y}_\ell$ ($2 \leq \ell \leq N - 1$) is associated with the input mode of the ℓth apparatus, while $\hat{Y}_N$ is related to the Nth outcome. The unitary operator $\hat{W}_\ell$ is determined by imposing that $\hat{K}_\ell$ is a positive operator. The identity $\sum_{n=1}^{2} \hat{M}_n^{(\ell)\dagger} \hat{M}_n^{(\ell)} = \hat{I}_2$ leads to the relation $\hat{K}_\ell^\dagger \hat{K}_\ell + \hat{Y}_{\ell+1}^\dagger \hat{Y}_{\ell+1} = \hat{Y}_\ell^\dagger \hat{Y}_\ell$. This relation indicates the conservation law of probability. Using this formula, we show that $\sum_{\ell=1}^{N} \hat{K}_\ell^\dagger \hat{K}_\ell = \hat{I}_2$.

A number of interesting quantum protocols with multi-outcome POVM's have been proposed in the literature. Two of the present authors (SA and FN)[14] proposed a measurement-only quantum feedback control of a single qubit, for example. Their proposal involves a four-outcome POVM satisfying $\text{Tr}\,\hat{K}_\ell^\dagger \hat{K}_\ell = 1/2$ and $\text{Tr}\,\hat{E}_\ell^\dagger \hat{E}_{\ell'} = f(x)$ for $\ell \neq \ell'$, with $\hat{E}_\ell = \hat{K}_\ell^\dagger \hat{K}_\ell$, a continuous real function f, and a real parameter x ($0 \leq x \leq 1$). We remark that f does not depend on the subscripts ℓ and ℓ', but is a function of x. The real variable x can be understood as the measurement strength. An important property of this POVM is that the indistinguishability is unbiased between arbitrary pairs of elements of the POVM. This mutually-unbiased feature can lead to an interesing quantum control. The present procedure is applicable to the construction of the corresponding measurement operators since one can freely control the number of outcomes, the trace of $\hat{K}_\ell^\dagger \hat{K}_\ell$, and the indistinguishability.

4. Solid-state qubits

Let us consider methods for implementing general weak measurements on solid-state qubits. In this paper, we focus on superconducting qubits.[35–38] Superconducting qubits have many advantages for quantum engineering. Their current experimental status [39] indicates that various important quantum operations, especially controlled operations are implemented reliably. The demonstration of controlled-NOT and controlled-phase gates was reported in various types of superconducting qubits.[39–45] Therefore, it is important for development of measurement-based quantum protocols to explore the systematic construction methods for general measurements in such interesting physical systems. Several theoretical studies on the implementation of general measurements in superconducting qubits have been reported in, e.g., Refs. 46–49.

Analogies with linear optical qubits are useful for designing measurement operators in superconducting qubits. Let us consider two superconducting qubits, one of which is the measured system, while the other is an ancillary system. The former corresponds to the polarization in the previous arguments, and the latter is regarded as the path degree of freedom in the interferometer setup. In the interferometer, the polarizing beam splitter plays a central role to create entanglement between the polarization and the path. This operation can be replaced with a controlled operation (e.g., a controlled-NOT gate) between the two superconducting qubits.

Now, we show a method for implementing a SASTOM on superconducting qubits. We use the following notation. The quantum states of the measured qubit is expressed in terms of the basis vectors $|+\rangle$ and $|-\rangle$ with $\langle +|-\rangle = 0$. The ancillary qubit is described by $|0\rangle$ and $|1\rangle$ with $\langle 0|1\rangle = 0$. First, we prepare an initial state in the total system

$$|\Psi^{(\text{in})}\rangle = |\psi\rangle \otimes (\alpha\,|0\rangle + \beta\,|1\rangle), \qquad (22)$$

with $\alpha^2 + \beta^2 = 1$, $\alpha, \beta \in \mathbb{R}$, and $0 \leq \alpha, \beta \leq 1$. We denote an arbitrary state in the measured qubit as $|\psi\rangle$. The state preparation in the ancillary system can be achieved using single-qubit operations. Next, we apply the controlled-NOT gate $|+\rangle\langle+| \otimes \hat{I}_2 + |-\rangle\langle-| \otimes \hat{\tau}_x$, where $\hat{\tau}_x = |0\rangle\langle1| + |1\rangle\langle0|$. The resultant state is $|\Psi^{(\text{out})}\rangle = \hat{M}_0\,|\psi\rangle \otimes |0\rangle + \hat{M}_1\,|\psi\rangle \otimes |1\rangle$, where using

$$\beta = \sqrt{1 - \alpha^2},$$

$$\hat{M}_0 = \alpha|+\rangle\langle+| + \sqrt{1 - \alpha^2}|-\rangle\langle-|,$$
$$\hat{M}_1 = \sqrt{1 - \alpha^2}|+\rangle\langle+| + \alpha|-\rangle\langle-|.$$

Therefore, by performing a projective measurement on the state of the ancillary qubit, we have a SASTOM on $|\psi\rangle$. The measurement direction can be changed using single-qubit gates on the measured system before the controlled-NOT gate.

The use of a partial controlled-NOT gate leads to the implementation of general two-outcome measurements. Let us now write down the recipe using $\hat{U} = |+\rangle\langle+| \otimes \hat{I}_2 + |-\rangle\langle-| \otimes \exp(i\xi\hat{\tau}_x)$ and the initial state (22). See, e.g., Ref. [50] for details of a theoretical proposal for performing $\hat{U}$. We find that $|\Psi^{(\text{out})}\rangle = \hat{U}|\Psi^{(\text{in})}\rangle = \hat{X}_0|\psi\rangle \otimes |0\rangle + \hat{X}_1|\psi\rangle \otimes |1\rangle$, where $\hat{X}_0 = \alpha|+\rangle\langle+| + (\alpha\cos\xi + i\beta\sin\xi)|-\rangle\langle-|$ and $\hat{X}_1 = \beta|+\rangle\langle+| + (i\alpha\sin\xi + \beta\cos\xi)|-\rangle\langle-|$. Depending on the readout result of the ancillary qubit, a proper single-qubit operation on the measured qubit is performed. Then, we find that the state $|\psi\rangle$ is transformed by the positive operator part of $\hat{X}_n$. Using the right-polar decomposition, we obtain the positive operator parts of $\hat{X}_0$ and $\hat{X}_1$, respectively,

$$\hat{M}_0 = \alpha|+\rangle\langle+| + \sqrt{1 - \alpha'^2}|-\rangle\langle-|,$$
$$\hat{M}_1 = \sqrt{1 - \alpha^2}|+\rangle\langle+| + \alpha'|-\rangle\langle-|,$$

where $\alpha' = \sqrt{[1 - (2\alpha^2 - 1)\cos(2\xi)]/2}$.

General measurements with multiple outcomes can be implemented in a similar manner to that given in Sec. 3.3. If we obtain the result 0 in the ancillary qubit, we do nothing. A measurement operator [i.e., $\hat{K}_1$ in Eq. (21)] is applied to $|\psi\rangle$. Otherwise we perform a single-qubit operation on the measured qubit to change the measurement direction and prepare a new superposition state in the ancillary qubit. Then, we apply a partial controlled-NOT gate to the two qubits again. Depending on the readout results of the ancillary qubit, we either obtain one element in the desired POVM [i.e., $\hat{K}_2$ in Eq. (20)] or continue to the next step. Repeating this procedure, we can obtain any POVM with multiple outcomes. Compared to linear optical qubits, the implementation of a general multi-outcome measurement in superconducting qubits has an advantage with respect to scalability. In linear optical qubits, it is necessary for the implementation of a general multi-outcome measurement to prepare all the optical elements corresponding to all the possible outcomes before the measurement. When the number of the outcomes is large, the setup become large and complicated. In addition, most of the elements in the measurement apparatus are irrelevant to the state in any single run. For example, if one obtains the outcome corresponding to $\hat{K}_1^\dagger \hat{K}_1$, the remaining parts of the measurement apparatus are not used. In superconducting qubits, the ancillary qubit can be used in the different steps of the measurement process. In contrast to linear optical setups, the total system is a two-qubit system even if the number of outcomes is large.

5. Summary

We have proposed methods for implementing general measurements on a single qubit in linear optical and solid-state qubits. We focused on three types of general measurements on $\mathbb{C}^2$. The first type is the SASTOM described by Eqs. (5) and (6). Their associated POVM is regarded as a minimal extension of a projection-valued measure. The second one is the general two-outcome measurements described by Eqs. (15) and (16). This is the most general form of the measurements with two outcomes on $\mathbb{C}^2$. These two kinds of measurements have only two outcomes. Finally, we found that the recursive construction given in Eq. (20) with general two-outcome measurements allows the design of general N-outcome measurements.

The studies on measurement in quantum mechanics provide an interesting research field for both fundamental physics and applications. Systematic and simple methods for the design of general measurements contributes to the development of this research area.

Acknowledgments

YO is partially supported by the Special Postdoctoral Researchers Program, RIKEN. S.A. and F.N. acknowledge partial support from the ARO, RIKEN iTHES project, JSPS-RFBR Contract No. 12-02-92100, Grant-in-Aid for Scientific Research (S), MEXT Kakenhi on Quantum Cybernetics, and Funding Program for Innovative R&D on S&T.

References

1. J. von Neumann, *Mathematical Foundations of Quantum Mechanics* (Princeton University Press, Princeton, 1955).
2. E. B. Davies, *Quantum Theory of Open Systems* (Academic Press, London, 1976).
3. K. Kraus, *States, Effects and Operations: Fundamental Notations of Quantum Theory* (Springer, Berlin, 1983).

4. A. Peres, *Quantum Theory: Concepts and Methods* (Kluwer Academic Publishers, Dordrecht, 1993).

5. V. B. Braginsky and F. Ya. Khalili, *Quantum measurement*, edited by K. S. Thorne (Cambridge University Press, Cambridge, England, 1992).

6. K. Banaszek, Phys. Rev. Lett. **86**, 1366 (2001).

7. A. A. Clerk, M. H. Devoret, S. M. Girvin, and R. J. Schoelkopf, Rev. Mod. Phys. **82**, 1155 (2010).

8. H. M. Wiseman and G. J. Milburn, *Quantum Measurement and Control* (Cambridge university press, Cambridge, England, 2010).

9. M. Koashi and M. Ueda, Phys. Rev. Lett. **82**, 2598 (1999).

10. H. Nakazato, T. Takazawa, and K. Yuasa, Phys. Rev. Lett. **90**, 060401 (2003).

11. J. K. Stockton, R. van Handel, and H. Mabuchi, Phys. Rev. A **70**, 022106 (2004).

12. A. Lupaşcu, S. Saito, T. Picot, P. C. de Groot, C. J. P. Harmans, and J. E. Mooij, Nature Phys. **3**, 119 (2007).

13. A. N. Korotkov and K. Keane, Phys. Rev. A **81**, 040103(R) (2010).

14. S. Ashhab and F. Nori, Phys. Rev. A **82**, 062103 (2010); H. M. Wiseman, Nature (London) **470**, 178 (2011).

15. Y. Ota, S. Ashhab, and F. Nori, J. Phys. A: Math. and Theore. **45**, 415303 (2012).

16. R. Inoue, S. Tanaka, R. Namiki, T. Sagawa, and T. Takahashi, Phys. Rev. Lett. **110**, 163602 (2013).

17. B. Huttner, A. Muller, J. D. Gautier, H. Zbinden, and N. Gisin, Phys. Rev. A **54**, 3783 (1996).

18. G. G. Gillett, R. B. Dalton, B. P. Lanyon, M. P. Almeida, M. Barbieri, G. J. Pryde, J. L. O'Brien, K. J. Resch, S. D. Bartlett, and A. G. White, Phys. Rev. Lett. **104**, 080503 (2010).

19. P. G. Kwiat, S. Barraza-Lopez, A. Stefanov, and N. Gisin, Nature (London) **409**, 1014 (2001).

20. Y. S. Kim, Y. W. Cho, Y. S. Ra, and Y. H. Kim, Opt. Express **17**, 11978 (2009).

21. N. Katz, M. Neeley, M. Ansmann, R. C. Bialczak, M. Hofheinz, E. Lucero, A. O'Connell, H. Wang, A. N. Cleland, J. M. Martinis, and A. N. Korotkov, Phys. Rev. Lett. **101**, 200401 (2008).

22. M. Iinuma, Y. Suzuki, G. Taguchi, Y. Kadoya, and F. Hofmann, New J. Phys. **13**, 033041 (2011).

23. S. Kocsis, B. Braverman, S. Raverts, M. J. Stevens, R. P. Mirin, L. K. Shalm, and A. M. Steinberg, Science **332**, 1170 (2011).

24. Y. Ota, S. Ashhab, and F. Nori, Phys. Rev. A **85**, 043808 (2012).

25. S. Sponar, J. Klepp, R. Loidl, S. Filipp, K. Durstberger-Rennhofer, R. A. Bertlmann, G. Badurek, H. Rauch, and Y. Hasegawa, Phys. Rev. A **81**, 042113 (2010).

26. J. B. Spring, B. J. Metcalf, P. C. Humphreys, W. S. Kolthammer, X.-M. Jin, M. Barbieri, A. Datta, N. Thomas-Peter, N. K. Langford, D. Kundys, J. C. Gates, B. J. Smith, P. G. R. Smith, I. A. Walmsley, Science **339**, 798 (2012).

27. J. L. van Hemmen, Z. Phys. B **38**, 271 (1980).

28. H. Umezawa, *Advanced Field Theory: Micro, Macro, and Thermal Physics* (AIP, New York, 1993).

29. R. Loudon, *The Quantum Theory of Light*, Third edition (Oxford Science Publications, New York, 2000).

30. P. Kok, W. J. Munro, K. Nemoto, T. C. Ralph, J. P. Dowling, and G. J. Milburn, Rev. Mod. Phys. **79**, 135 (2007).

31. R. A. Horn and C. R. Johnson, *Matrix Analysis* (Cambridge University Press, Cambridge, England, 1985) Chap.7.

32. D. Meschede, *Optics, Light and Lasers: The Practical Approach to Modern Aspects of Photonics and Laser Physics* Seconde, Revised and Enlarged Edition (Wiley-VCH, Weinheim, 2007) Chap.3.

33. R. Simon and N. Mukunda, Phys. Lett. A **143**, 165 (1990).

34. R. Bhandari and T. Dasgupta, Phys. Lett. A **143**, 170 (1990).

35. J. Q. You and F. Nori, Phys. Today **58** (11), 42 (2005).

36. M. Nakahara and T. Ohmi, *Quantum Computing: From Linear Algebra to Physical Realizations* (Taylor & Francis, Boca Raton, FL, 2008) Chap.15.

37. J. Clarke and F. K. Wilhelm, Nature (London) **453**, 1031 (2008).

38. J. Q. You and F. Nori, Nature (London) **474**, 589 (2011).

39. I. Buluta, S. Ashhab, and F. Nori, Rep. Prog. Phys. **74**, 104401 (2011).

40. T. Yamamoto, Yu. A. Pashkin, O. Astafiev, Y. Nakamura, and J. S. Tsai, Nature **425**, 941 (2003).

41. J. H. Plantenberg, P. C. de Groot, C. J. P. M. Harmans, and J. E. Mooij, Nature **447**, 836 (2007).

42. L. DiCarlo, J. M. Chow, J. M. Gambetta, Lev S. Bishop, B. R. Johnson, D. I. Schuster, J. Majer, A. Blais, L. Frunzio, S. M. Girvin, and R. J. Schoelkopf, Nature **460**, 240 (2009).

43. M. Neeley, R. C. Bialczak, M. Lenander, E. Lucero, M. Mariantoni, A. D. O'Connell, D. Sank, H. Wang, M. Weides, J. Wenner, Y. Yin, T. Yamamoto, A. N. Cleland, and J. M. Martinis, Nature **467**, 570 (2010).

44. P. C. de Groot, J. Lisenfeld, R. N. Schouten, S. Ashhab, A. Lupascu, C. J. P. M. Harmans, and J. E. Mooij, Nature Phys. **6**, 763 (2010).

45. J. M. Chow, A. D. Córcoles, J. M. Gambetta, C. Rigetti, B. R. Johnson, J. A. Smolin, J. R. Rozen, G. A. Keefe, M. B. Rothwell, M. B. Ketchen, and M. Steffen, Phys. Rev. Lett. **107**, 080502 (2011).

46. A. N. Korotkov and A. N. Jordan, Phys. Rev. Lett. **97**, 166805 (2006).

47. G. S. Paraoanu, Europhys. Lett. **93**, 64002 (2011).

48. G. S. Paraoanu, Found. Phys. **41**, 1214 (2011).

49. S. Ashhab, J. Q. You, and F. Nori, Phys. Rev. A **79**, 032317 (2009); New J. Phys. **11**, 083017 (2009); Phys. Scr. **T137**, 014005 (2009).

50. P. C. de Groot, S. Ashhab, A. Lupascu, L. DiCarlo, F. Nori, C. J. P. M. Harmans, and J. E. Mooij, New J. Phys. **14**, 073038 (2012).

FAST AND ACCURATE SIMULATION OF QUANTUM COMPUTING BY MULTI-PRECISION MPS: RECENT DEVELOPMENT

AKIRA SAITOH*

Quantum Information Science Theory Group, National Institute of Informatics, 2-1-2 Hitotsubashi, Chiyoda, Tokyo 101-8430, Japan

The time-dependent matrix-product-state (TDMPS) simulation method has been known as one of fast simulation methods to study time-evolving quantum systems. Here, I report recent development of my open-source C++ library named ZKCM_QC designed for TDMPS simulations of quantum circuits with arbitrary floating-point precision. Simulation performance is reported for well-known quantum algorithms. In addition, it is numerically shown that a trustworthy simulation should be performed in multiprecision and should not involve truncations of nonzero Schmidt coefficients.

Keywords: Time-dependent matrix product states; Quantum computing

1. Introduction

Classical quantum-circuit simulators[1–9] are practical tools to study quantum computing[10] for the time being as it is quite far beneath the stage of production. In this regard, the time-dependent matrix-product-state (TDMPS) method is a useful simulation method, which was introduced by Vidal in 2003,[3] for a fast simulation of quantum computing when it does not involve a large amount of entanglement. It simulates a quantum circuit within the cost of $O(q_g m_{\mathrm{max,max}}^3)$ where q_g is the number of single-qubit and two-qubit operations in the circuit; $m_{\mathrm{max,max}} := \max_{s,t} m(s,t)$ with $m(s,t)$ the Schmidt rank for the splitting between the sites s and $s+1$ at time t. Thus, a polynomial time simulation is possible if the Schmidt rank grows only polynomially in the input size.

The TDMPS method has been, however, used mainly for evaluating the time dependence of physical properties of condensed matters[11,12] rather

---

*Present address: Department of Computer Science and Engineering, Toyohashi University of Technology, 1-1 Hibarigaoka, Tenpaku-cho, Toyohashi, Aichi 441-8580, Japan.

than simulating quantum algorithms in the physics community as far as the author knows. There have been a few works on TDMPS simulations of quantum algorithms: Kawaguchi et al.[13] simulated the Grover search for a simple oracle and showed that the simulation cost was polynomial in the number of qubits. This was because of the simple oracle structure. Later I simulated a variant of the Brüschweiler search and showed that all the solutions could be found within polynomial time when the oracle structure was simple enough.[5] Recently, Chamon and Mucciolo[14] theoretically showed that an integer computation based on TDMPS could solve a search problem within a subexponential time (i.e. $O(2^{n^c})$ time with $c < 1$, which is smaller than the query complexity of the Grover search) as long as the oracle circuit can be decomposed to less than $O(n^2)$ two-qubit gates. Besides, Bañuls et al.[15] used a TDMPS simulation of an adiabatic time evolution for solving an exact-cover SAT problem.[16] They reported that they could simulate a time evolution of a 100-qubit system with the threshold 14 for the Schmidt rank.

Thus, there have been several evidences for the usefulness of TDMPS for fast simulation of quantum algorithms although there have not been many authors working in this direction. It is easily expected that more researchers will have an interest if there are user-friendly free softwares for this purpose. One choice is the well-known ALPS package,[17] which is a general-purpose simulation library for condensed matter physics. It has a routine for TDMPS but it can only be used for simulating an adiabatic time evolution under given initial and final Hamiltonians. The other choice is my C++ library ZKCM_QC,[18] which is an extension library of the ZKCM library[9] developed with an emphasis of an easy-to-use syntax for multiprecision matrix computation. It uses GMP[19] and MPFR[20] as back-end libraries for multiprecision floating-point computation so that the outputs of ZKCM_QC are accurate for more than several tens of qubits for which double-precision computation causes significant rounding errors during simulation.

In this report, we firstly revisit the basics of the TDMPS method in Sec. 2. The demand of multiprecision computation is briefly described in Sec. 3. Actual simulation performance of the ZKCM_QC library is reported in Sec. 4 in which the standard quantum algorithms, namely, the Deutsch-Jozsa algorithm,[21] the Grover search,[22] and the Shor's algorithm[23] are simulated. Simulation results of an in-place addition is also explained separately since this is one of the important components for economical implementation of the Shor's algorithm. Concluding remarks are given in Sec. 5.

2. Basics of the TDMPS method

Here, we begin with a convention of notations. The computational basis is represented as $\{|0\rangle, |1\rangle\}^n$ for n-qubit quantum states with $|0\rangle = \begin{pmatrix} 1 & 0 \end{pmatrix}^{\mathrm{T}}$ and $|1\rangle = \begin{pmatrix} 0 & 1 \end{pmatrix}^{\mathrm{T}}$. An n-qubit quantum state is represented as $|\Psi\rangle = \sum_{i_0 \cdots i_{n-1}=0 \cdots 0}^{1 \cdots 1} c_{i_0 \cdots i_{n-1}} |i_0 \cdots i_{n-1}\rangle$ with complex amplitudes $c_{i_0 \cdots i_{n-1}}$. We employ the Vidal's MPS form[3,5] for our TDMPS simulations:

$$
\begin{aligned}
|\Psi\rangle = \sum_{i_0 \cdots i_{n-1}=0 \cdots 0}^{1 \cdots 1} \Bigg[&\sum_{v_0=0}^{m_0-1} \sum_{v_1=0}^{m_1-1} \cdots \sum_{v_{n-2}=0}^{m_{n-2}-1} Q_0(i_0, v_0) V_0(v_0) \\
&\times Q_1(i_1, v_0, v_1) \cdots Q_s(i_s, v_{s-1}, v_s) V_s(v_s) Q_{s+1}(i_{s+1}, v_s, v_{s+1}) \cdots \quad (1) \\
&\cdots V_{n-2}(v_{n-2}) Q_{n-1}(i_{n-1}, v_{n-2}) \Bigg] |i_0 \cdots i_{n-1}\rangle,
\end{aligned}
$$

where we use tensors $\{Q_s\}_{s=0}^{n-1}$ with parameters i_s, v_{s-1}, v_s (v_{-1} and v_{n-1} are excluded) and $\{V_s\}_{s=0}^{n-2}$ with parameter v_s. In addition, tensor $V_s(v_s)$ stores the Schmidt coefficients for the splitting between the sth site and the $(s+1)$th site; m_s is a suitable number of Schmidt coefficients which does not exceed a threshold m_{trunc} of one's choice.

For example, $|0 \cdots 0\rangle$ is represented in this form by setting all m_s to 1, and setting all $Q_s(0, 0, 0)$ and $V_s(0)$ to 1, and $Q_s(1, 0, 0)$ to 0.

Under the MPS representation, a unitary time evolution can be computed by taking the corresponding space only into account, $i.e.$, we have only to handle the tensors for the space of our concern. For example, consider a unitary operation $U = \sum_{k,k'=0}^{1} U_{kk'} |k\rangle\langle k'|$ acting on the qubit s. Then the resultant state is computed by updating Q_s in the following way. $Q_s(i_s, v_{s-1}, v_s) \overset{U}{\mapsto} \widetilde{Q}_s(i_s, v_{s-1}, v_s)$ with $\widetilde{Q}_s(i_s, v_{s-1}, v_s) = \langle i_s| \sum_{k,k'} U_{k,k'} Q_s(k', v_{s-1}, v_s) |k\rangle$.

Time evolution under a two-qubit unitary operation acting on qubits s and $s+1$, $U = \sum_{k_s k_{s+1}, k'_s k'_{s+1}} U_{(k_s k_{s+1})(k'_s k'_{s+1})} |k_s\rangle |k_{s+1}\rangle \langle k'_s| \langle k'_{s+1}|$, can also be computed in a similar manner with a little complicated process. This process updates the tensors Q_s, V_s, and Q_{s+1}. Let us firstly write the state in the following way.

$$
\begin{aligned}
|\Psi\rangle = \sum_{v_{s-1}=0}^{m_{s-1}} \sum_{i_s=0}^{1} \sum_{i_{s+1}=0}^{1} \sum_{v_{s+1}=0}^{m_{s+1}} \Bigg[&V_{s-1}(v_{s-1}) |v_{s-1}\rangle \\
&\otimes W(i_s, i_{s+1}, v_{s-1}, v_{s+1}) |i_s\rangle |i_{s+1}\rangle \otimes V_{s+1}(v_{s+1}) |v_{s+1}\rangle \Bigg]
\end{aligned}
$$

with $|v_{s-1}\rangle$ the left Schmidt vectors for the splitting between sites $s-1$ and s, $W(i_s, i_{s+1}, v_{s-1}, v_{s+1}) = \sum_{v_s} Q_s(i_s, v_{s-1}, v_s) V_s(v_s) Q_{s+1}(i_{s+1}, v_s, v_{s+1})$, and $|v_{s+1}\rangle$ the right Schmidt vectors for the splitting between sites $s+1$

and $s + 2$. The unitary transformation is applied to the tensor W in the following way. $W(i_s, i_{s+1}, v_{s-1}, v_{s+1}) \overset{U}{\mapsto} \widetilde{W}(i_s, i_{s+1}, v_{s-1}, v_{s+1})$ with

$$\widetilde{W}(i_s, i_{s+1}, v_{s-1}, v_{s+1}) =$$

$$\langle i_s|\langle i_{s+1}| \sum_{k_s k_{s+1}, k'_s k'_{s+1}} U_{(k_s k_{s+1})(k'_s k'_{s+1})} W(k'_s, k'_{s+1}, v_{s-1}, v_{s+1})|k_s\rangle|k_{s+1}\rangle.$$

Now the resultant state is written as

$$|\widetilde{\Psi}\rangle = \sum_a \sum_b R_{ab}|a\rangle|b\rangle$$

with labels $a = (v_{s-1} i_s)$ and $b = (i_{s+1} v_{s+1})$, and matrix R whose (a, b) element is $R_{ab} = V_{s-1}(v_{s-1})\widetilde{W}(i_s, i_{s+1}, v_{s-1}, v_{s+1})V_{s+1}(v_{s+1})$. Then, we perform a singular value decomposition (SVD) of R. This results in $R = ADB^\dagger$ where A is a $2m_{s-1} \times 2m_{s-1}$ unitary matrix, D is a $2m_{s-1} \times 2m_{s+1}$ matrix with only diagonal elements in the upper-left side, and B is a $2m_{s+1} \times 2m_{s+1}$ unitary matrix. This SVD is performed in the way that the singular values are found in the descending order in D. Suppose there are q nonvanishing singular values. This SVD can also be written as $R_{ab} = \sum_{d=0}^{q-1} A_{ad} D_{dd}(B_{bd})^*$. We choose at most m_{trunc} elements among D_{dd}'s (from larger to smaller) and store them into the tensor $\widetilde{V}_s(v_s)$. Hence, $\widetilde{m}_s = \min(q, m_{\text{trunc}})$ elements are stored. Then, we have

$$|\widetilde{\Psi}\rangle = \sum_{v_s=0}^{\widetilde{m}_s - 1} \widetilde{V}_s(v_s)|l(v_s)\rangle|r(v_s)\rangle,$$

where $|l(v_s)\rangle = \sum_a A_{av_s}|a\rangle$ and $|r(v_s)\rangle = \sum_b (B_{bv_s})^*|b\rangle$. Note that $|l(v_s)\rangle$ and $|r(v_s)\rangle$ are represented in the basis $|v_{s-1}\rangle|i_s\rangle$ and the basis $|i_{s+1}\rangle|v_{s+1}\rangle$, respectively. By writing the basis labels explicitly, we have $|l(v_s)\rangle = \widetilde{Q}_s(i_s, v_{s-1}, v_s)|v_{s-1}\rangle|i_s\rangle$ and $|r(v_s)\rangle = \widetilde{Q}_{s+1}(i_{s+1}, v_s, v_{s+1})|i_{s+1}\rangle|v_{s+1}\rangle$. In this way, the tensors Q_s, V_s, and Q_{s+1} are updated to $\widetilde{Q}_s$, $\widetilde{V}_s$, and $\widetilde{Q}_{s+1}$.

It is well-known that single-qubit and two-qubit operations are sufficient for performing universal quantum computation. The ZKCM_QC library, however, uses three-qubit operations as basic operations in order to avoid an overhead in a circuit construction. Simulation of a three-qubit quantum gate acting on consecutive qubits requires an update of tensors Q_s, V_s, Q_{s+1}, V_{s+1}, and Q_{s+2}. This is a more complicated process than the above-described one. For the details, see the appendix of Ref. 9.

It is also possible to simulate a single-qubit projective measurement. Consider a projection operation P acting on the sth qubit. First we update the tensor Q_s to $\widetilde{Q}_s$ with $\widetilde{Q}_s(i_s, v_{s-1}, v_s) = \langle i_s| \sum_{i'_s} Q_s(i'_s, v_{s-1}, v_s)P|i'_s\rangle/$

$\sqrt{\mu}$ with $\mu = \| \sum_{i_s, v_{s-1}, v_s} Q_s(i_s, v_{s-1}, v_s) P |i_s \rangle \|^2$. Then, we need to update the tensors corresponding to the other qubits as long as they are correlated with the sth qubit. This can be done by sequentially using the same process as applying $I \otimes I$ to the consecutive qubits, moving the *cursor* from s to 0 and also from s to $n - 1$. We need not to update the tensors beyond the place where the Schmidt rank is one.

2.1. *Operator-space TDMPS*

Although I do not use the operator-space TDMPS for simulations presented in this contribution, it is beneficial to mention about it as it is quite straight-forward to migrate from the standard TDMPS. Here, we follow the formulation by Zwolak.[24] It is used for simulating time evolution of a density operator under trace-preserving completely-positive (TPCP) maps.

The few things we should pay attention to are the computational basis and the definition of the inner product. As for the basis for a qubit, we need to employ an operator basis, typically $\{|0\rangle, |1\rangle, |2\rangle, |3\rangle\}$ where $|0\rangle = I$, $|1\rangle = X$, $|2\rangle = Y$, and $|3\rangle = Z$ are standard Pauli matrices [here, $Z = \mathrm{diag}(1, -1)$]. As for the inner product, its definition should be given as $(A|B) = \mathrm{Tr}(A^\dagger B)/d$ for operators A and B acting on a d-dimensional Hilbert space. Then the operator MPS representation of an n-qubit density matrix is

$$|\rho) = \sum_{i_0 \cdots i_{n-1}=0 \cdots 0}^{3 \cdots 3} \left[\sum_{v_0=0}^{m_0-1} \sum_{v_1=0}^{m_1-1} \cdots \sum_{v_{n-2}=0}^{m_{n-2}-1} Q_0(i_0, v_0) V_0(v_0) \right.$$
$$\times Q_1(i_1, v_0, v_1) \cdots Q_s(i_s, v_{s-1}, v_s) V_s(v_s) Q_{s+1}(i_{s+1}, v_s, v_{s+1}) \cdots \quad (2)$$
$$\left. \cdots V_{n-2}(v_{n-2}) Q_{n-1}(i_{n-1}, v_{n-2}) \right] |i_0 \cdots i_{n-1}).$$

This is quite similar to the MPS of a pure state we have seen in Eq. (1). For a simple example, $(I/2)^{\otimes n}$ is represented by an operator MPS with tensor data $Q_0(0, 0) = 1/2$, $Q_s(0, 0, 0) = 1/2$ ($s = 1, \ldots, n - 2$), $Q_{n-1}(0, 0) = 1/2$, and $V_s(0) = 1$ ($s = 0, \ldots, n - 2$).

Time evolution of $|\rho)$ caused by a TPCP map Λ can be simulated in the same manner as we simulate a unitary time evolution in the standard TDMPS method. Here, Λ must be represented as a square matrix with the $|k)(k'|$ notation, *i.e.*, $\Lambda = \sum_{kk'} \Lambda_{kk'} |k)(k'|$. For example, time evolution caused by a TPCP map Λ acting on the sth qubit is simulated by updating $Q_s(i_s, v_{s-1}, v_s)$ to $\widetilde{Q}_s(i_s, v_{s-1}, v_s) = (i_s| \sum_{k=0, k'=0}^{3,3} \Lambda_{k,k'} Q_s(k', v_{s-1}, v_s) |k)$. Time evolution caused by a two-qubit TPCP map can also be simulated in the same manner as we have already seen for the two-qubit unitary gate simulation in the standard TDMPS method.

It should be useful to mention that a unitary operation U acting on a d-dimensional Hilbert space can be easily translated into a map $\mathcal{U}$ acting on a d^2-dimensional operator Hilbert space. Let us denote the operator basis operators as $|k\rangle \equiv |\sigma_k\rangle$ with σ_k the kth basis operator. Then, the (i, j) element of the matrix representation of $\mathcal{U}$ is $\mathcal{U}_{ij} = (i|\mathcal{U}|j) = \mathrm{Tr}(\sigma_i U \sigma_j U^\dagger)/d$.

3. Necessity of multiprecision computation

Multiprecision computation has been utilized in computational physics to avoid the accumulation of rounding errors in sensitive simulations of dynamics,[25] but has not been widely used in the community. For the TDMPS simulation of quantum circuits, I demonstrated that multiprecision computation is requisite in order for avoiding a significant error in the simulation results.[18]

The particular findings in Ref. 18 were as follows. In both cases, a truncation of nonzero Schmidt coefficients was not employed.

(i) A TDMPS simulation of a five-qubit circuit for a three-qubit Grover search was performed. In the Grover routine R_{G}, the oracle part was set to $U_{\mathrm{o}} = 1 - 2|101\rangle\langle 101|$. Thus, R_{G} was set to $(1 - 2|s\rangle\langle s|)U_{\mathrm{o}}$ with $|s\rangle = (1/\sqrt{8})\sum_{x_0 x_1 x_2=000}^{111} |x_0 x_1 x_2\rangle$. The exact probability of finding the solution after twenty Grover iterations, $p_{20} = |\langle 101|R_{\mathrm{G}}^{20}|s\rangle|^2$, was calculated exactly by a symbolic computation. The error in the probability $\widetilde{p_{20}}$ computed by the simulation was evaluated as the quantity $E = |\widetilde{p_{20}} - p_{20}|$. This numerical error E was plotted against the floating point precision employed for the simulation. It was found that E was larger than 0.035 and did not change largely until the precision was enhanced beyond the double precision. It suddenly dropped around the 55-bits precision and almost vanished for more than 70-bits precision.

(ii) An n qubit circuit performing $(\mathrm{QFT}^{-1})(\mathrm{QFT})(\mathrm{CNOT}_{0,n-1})$ $H_0|0_0 \cdots 0_{n-1}\rangle$ was considered, where QFT is a quantum Fourier transform, $\mathrm{CNOT}_{0,n-1}$ is a controlled NOT gate acting on the 0th and $(n-1)$th qubits, and H_0 is the Hadamard transform acting on the 0th qubit. Since $(\mathrm{QFT}^{-1})(\mathrm{QFT})$ is just an identity map, the resultant state should be $(|0_0 0_{n-1}\rangle + |1_0 1_{n-1}\rangle)/\sqrt{2} \otimes |0_1 \cdots 0_{n-2}\rangle$. In a TDMPS simulation, however, there is a numerical error to some extent in the computed resultant state. The error was quantified by $E = |\langle 00|\widetilde{\rho'}|11\rangle - 1/2|$ with $\widetilde{\rho'}$ the computed reduced density matrix of the 0th and the $(n-1)$th qubits of the resultant state. For $n = 8, 14$, and 20, E was plotted against the floating point precision. It was found that, for all of these values of n, a sudden drop of E was observed.

These simulation results suggest that there are cases where a numerical error is significant until we go beyond a certain threshold for the floating point precision. It is thus recommended to see the behavior of computational results as functions of the floating point precision in TDMPS simulations of quantum circuits.

4. Performance

As mentioned, my simulation library ZKCM_QC uses multiprecision floating-point operations provided by GMP[19] and MPFR[20] for basic operations. Thus the basic operations are inevitably slow in comparison to fixed-precision floating-point operations. The way to improve the performance by the author's effort is therefore limited to choosing algorithms for matrix manipulations carefully and making technical elaborations. Here, I report the recent progress in this regard. Firstly, an effort of speeding-up the routine for Hermitian-matrix diagonalization will be reported in subsection 4.1. Then, the simulation performance of my library will be reported for typical quantum algorithms in the remaining subsections. More specifically, results for the Deutsch-Jozsa algorithm, the Grover search, an in-place addition, and the Shor's prime factorization are explained in subsections 4.2, 4.3, 4.4, and 4.5, respectively.

4.1. *Recent speedup in Hermitian-matrix diagonalization*

Speed of Hermitian matrix diagonalization is a large factor of the actual speed of the TDMPS simulation using ZKCM_QC since it uses the routine of Hermitian-matrix diagonalization for the singular value decomposition. As of version 0.3.6 of ZKCM, the speed has been improved significantly and now that it is faster than that of the famous PARI library.[26] Our routine uses the standard Householder-QR method for Hermitian matrices[27] and is named "diag_H".

It should be noted that conventional multiprecision routines for diagonalization have not been useful for our purpose. The routine "eigen" of PARI does not work for degenerate subspaces (see the PARI/GP bug report logs - #1349, August 2012). Thus it cannot be used in TDMPS simulations of quantum circuits, since most of the subspaces we handle are degenerate (this situation is quite common whenever we have Hadamard gates and CNOT gates). PARI also has another routine "jacobi" but this works for real symmetric matrices only. Thus a workaround is needed to use it for our purpose. [For an Hermitian matrix A, the workaround is to use a symmetric matrix $\begin{pmatrix} \text{Re}(A) & -\text{Im}(A) \\ \text{Im}(A) & \text{Re}(A) \end{pmatrix}$ (see Ch. 11.5 of Ref. 28).] In addition, it uses the Jacobi's method so that it is slower than the House-holder's method. Furthermore, we need to double the precision of an input matrix when "jacobi" is used together with this workaround, in order to support specified precision as far as I tested. Another routine is found in the multiprecision LAPACK.[29] Nevertheless, this library cannot be used for the back-end of TDMPS because it has a serious bug in the matrix diagonalization (see a bug report on 26 July 2012 in the Mplapack-devel mailing list). It fails for computing eigenvalues for some matrices with relatively large corner elements and some vanishing center elements (typically a density matrix of an entangled state); this bug has been unfixed yet.

Here I compare the routine "diag_H" of ZKCM versions 0.3.6 and 0.3.2 with the routines "eigen" and "jacobi" of PARI version 2.5.3. For this comparison, I test the average time consumption to find all the eigenvectors of a random $N \times N$ Hermitian matrix with a unit Frobenius norm for precision prec [bits].

First I set N to 100 and tried several values of precision (Table 1). The time consumptions of the routines were on the same order of magnitude for the tested precision between 256 and 1280 [bits]. "diag_H" of ZKCM version 0.3.6 was the fastest among the compared routines. Second, I fixed the precision to 768 [bits] and tried several values of N. As shown in Table 2, for the tested values between 25 and 125, the time consumptions were again on the same order of magnitude, and "diag_H" of ZKCM version 0.3.6 was the fastest for $N \geq 75$.

4.2. Simulation of the Deutsch-Jozsa algorithm

Here we will see a TDMPS simulation of the Deutsch-Jozsa algorithm.[21] In a brief explanation, the problem instance is a function $f : \{0, 1\}^l \to \{0, 1\}$ that is either balanced or constant. [Note: f is balanced if $\#\{\mathbf{x}|f(\mathbf{x}) = 0\} = \#\{\mathbf{x}|f(\mathbf{x}) = 1\}$ where $\mathbf{x} \in \{0_0 \cdots 0_{l-1}, \ldots, 1_0 \cdots 1_{l-1}\}$; f is constant if $f(\mathbf{x})$ is same for all $\mathbf{x}$.] The question is to decide whether f is balanced or constant. This takes $1 + 2^{l-1}$ queries for the worst case in classical computation. In contrast, it takes only a single query in quantum computation using the Deutsch-Jozsa algorithm. A sketch of the algorithm is as follows. (i) We apply $H^{\otimes l} V_f H^{\otimes l}$ to the l-qubit state $|0_0 \cdots 0_{l-1}\rangle$, where V_f is an operation mapping each $|\mathbf{x}\rangle$ to $(-1)^{f(\mathbf{x})}|\mathbf{x}\rangle$.

Table 1. Comparison of the average real time consumption for the program routines to find all the eigenvectors of a normalized random 100×100 Hermitian matrix. The average was taken over ten different matrices. The standard deviation is shown in parentheses in small fonts. "prec" stands for the precision. [Precision was doubled for "jacobi" (see the text)]. The programs were run as a single thread on a machine with an Intel Core i5 M460 2.53GHz CPU, 4GB memory, and the Fedora 15 64-bit OS. Note: For precision 256 [bits], function "eigen" of PARI stopped with an error and output no result.

prec	ZKCM 0.3.6, diag_H [sec]	ZKCM 0.3.2, diag_H [sec]	PARI, eigen [sec]	PARI, jacobi [sec]
256	73.0 (0.525)	175 (0.105)	N/A (N/A)	103 (0.324)
512	109 (0.624)	259 (0.160)	171 (1.27)	265 (0.974)
768	171 (0.259)	413 (0.378)	237 (1.24)	477 (2.10)
1024	276 (1.68)	632 (0.358)	378 (1.58)	726 (2.24)
1280	394 (2.18)	903 (0.315)	503 (1.21)	1020 (5.47)

Table 2. Comparison of the average real time consumption for the program routines to find all the eigenvectors of a normalized random $N \times N$ Hermitian matrix under the fixed precision 768 [bits] [precision was doubled for "jacobi" (see the text)]. The average was taken over ten different matrices. The standard deviation is shown in parentheses in small fonts. The programs were run as a single thread on a machine with an Intel Core i5 M460 2.53GHz CPU, 4GB memory, and the Fedora 15 64-bit OS.

N	ZKCM 0.3.6, diag_H [sec]	ZKCM 0.3.2, diag_H [sec]	PARI, eigen [sec]	PARI, jacobi [sec]
25	1.66 (0.0142)	3.41 (0.0152)	0.941 (0.00406)	5.99 (0.0699)
50	15.3 (0.0530)	34.5 (0.0624)	14.5 (0.0248)	50.9 (0.364)
75	61.2 (0.170)	146 (0.238)	74.3 (0.636)	182 (0.977)
100	171 (0.259)	413 (0.378)	237 (1.24)	477 (2.10)
125	386 (0.909)	961 (1.10)	596 (1.72)	1070 (7.40)

(ii) We measure the l qubits in the computational basis. The probability of finding the qubits in 0's simultaneously in this measurement vanishes when f is balanced; in contrast, it is exactly unity when f is constant.

More details of the algorithm are found in, *e.g.*, Sec. 3.1.2 of Ref. 10.

Here, let us consider a particular function $f(\mathbf{y}_0 \cdots \mathbf{y}_{N_g-1}) = \bigoplus_{i=0}^{N_g-1} g(\mathbf{y}_i)$ with $g(x_0 x_1 x_2 x_3) = (x_0 \wedge x_1) \vee (x_1 \wedge x_2) \vee (x_2 \wedge x_3)$ where $\mathbf{y}_i \in \{0,1\}^4$ and $x_j \in \{0,1\}$; N_g is a positive integer (symbol $\bigoplus$ stands for the exclusive OR operation). In Fig. 1, the quantum circuit of the algorithm for this function is depicted. This function is a balanced function for any $N_g \geq 1$. By the structure of the circuit, each of the N_g measurements should results in $\mathrm{Prob}(0000) = 0$ if there is no numerical error during simulation. This fact is easily proved: Assume that we have different values of $\mathrm{Prob}(0000)$ for two different bundles of qubits in the output. This contradicts to the

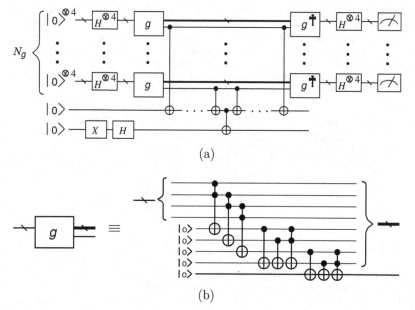

Fig. 1. (a) Quantum circuit of the Deutsch-Jozsa algorithm for the specified function (see the text). (b) Internal structure of gate g.

fact that the bundles are equivalent to each other by the circuit structure. Thus the assumption is denied.

In my previous contribution,[9] I employed $N_g = 7$ (namely, 65 qubits in total) and showed that truncation of even a single nonzero Schmidt coefficient caused a significant error in the computed value of Prob(0000). This was because none of nonzero Schmidt coefficients was negligible.

Now I show in this report how the simulation running time and $m_{\mathrm{max,max}}$ grows as the number n of qubits grows. As shown in Fig. 2, time consumption seems to be on the order of n^3 although further investigation is required. This is probably because the value of $m_{\mathrm{max,max}}$ became invariant for n larger than a certain value as shown in the figure. This phenomenon appeared probably because the quantum circuit was highly structured.

4.3. Simulation of the Grover search

In this subsection, the simulation performance of the ZKCM_QC library is evaluated for an example of the Grover's quantum search.[22] It was shown by Kawaguchi et al.[13] that the Grover search can be simulated efficiently with TDMPS if a simple oracle circuit is chosen (see also my TDMPS sim-

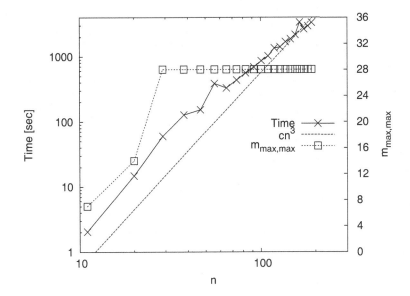

Fig. 2. Plots of running time and $m_{\text{max,max}}$ as functions of the number of qubits, $n = 9N_g + 2$. The line of cn^3 with $c = 5.58 \times 10^{-4}$ is also shown (using the left vertical axis). Non-averaged raw data were used for the data points. The program was run as a single thread on a machine with four Intel Xeon E7-8837 2.67GHz CPUs, 315GB memory, and the RedHat Enterprise Linux 6 64-bit OS.

ulation of a bulk-ensemble database search[5]). Here we consider the Grover search to solve a 3SAT[16] instance. 3SAT is a problem to decide if there is an assignment to the variables $x_0, \cdots, x_{v-1}$ such that a given conjunctive normal form (CNF) $\phi(x_0, \cdots, x_{v-1}) = C_0 \wedge \cdots \wedge C_{c-1}$ becomes true (namely, 1), where C_j is a clause $(l_\alpha \vee l_\beta \vee l_\gamma)$ with some three literals. Here, a literal l_τ is a variable x_τ or its logical negation $\neg x_\tau$. A CNF is given without ambiguity, $i.e.$, it is written in terms of $x_0, \cdots, x_{v-1}$ and logical operations.

Unlike the true Grover search for a quantum computer, a TDMPS simulation of the Grover search has only to handle a single query when an instance of a decision problem is given. This is because a slight change occurs in the polarization of the oracle qubit after a single query if and only if the given instance has a truth assignment among the 2^n possible assignments. Figure 3 describes the construction of an oracle quantum circuit for a 3SAT instance.

Consider the following instance: $\phi(x_0, \cdots, x_6) = (\neg x_2 \vee x_4 \vee x_6) \wedge (x_0 \vee x_1 \vee \neg x_3) \wedge (\neg x_1 \vee x_2 \vee \neg x_5) \wedge (x_0 \vee x_1 \vee x_4) \wedge (\neg x_0 \vee \neg x_4 \vee x_6) \wedge (x_2 \vee x_3 \vee \neg x_6) \wedge$

60

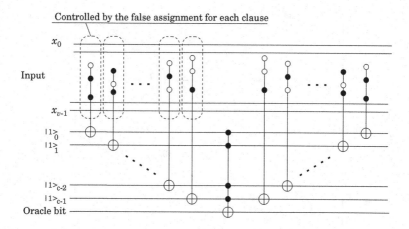

Fig. 3.   Quantum oracle circuit for a 3SAT instance with v variables and c clauses.

$(\neg x_2 \vee \neg x_5 \vee \neg x_6) \wedge (x_1 \vee \neg x_2 \vee \neg x_3) \wedge (x_1 \vee \neg x_3 \vee x_5) \wedge (x_1 \vee x_4 \vee \neg x_6)$, which is satisfiable. For this instance, a TDMPS simulation of a single query of the Grover search was performed. The program was run as a single thread on the machine with four Intel Xeon E7-8837 2.67GHz CPUs, 315GB memory, and the RedHat Enterprise Linux 6 64-bit OS. It took 132 minutes to perform this simulation when the floating-point precision was 412 [bits]. This is extremely expensive since any classical random seek may solve it within one millisecond. A possible reason of time consumption is a rather large maximum Schmidt rank. It was 40 and none of the nonzero Schmidt coefficients was negligible (Fig. 4).

4.4. Simulation of a QFT-based in-place addition

The in-place arithmetic circuits[30] using quantum Fourier transform (QFT) are quite often used for economical construction of quantum circuits as they do not require ancillary qubits. It is theoretically easily shown that a TDMPS simulation of a QFT-based addition circuit is fast in the sense that the Schmidt rank during QFT does not exceed the degree of superposition of the input state represented in the computational basis. Here, some theoretical explanation is given and a TDMPS simulation of a quantum circuit involving a simple QFT-based adder is performed.

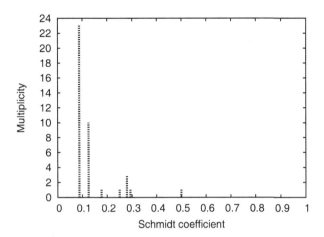

Fig. 4. Distribution of nonzero Schmidt coefficients at the point the Schmidt rank reached 40.

As is well-known in this community, QFT is defined by

$$\text{QFT} : |a\rangle \mapsto \frac{1}{\sqrt{2^n}} \sum_{l=0}^{2^n-1} e^{i2\pi la/2^n} |l\rangle$$

for an n-bit unsigned integer $a = a_{n-1}a_{n-2}\cdots a_0$. The resultant state is known to be separable:

$$\frac{1}{\sqrt{2^n}} \sum_{l=0}^{2^n-1} e^{i2\pi la/2^n} |l\rangle = |\phi_0(a)\rangle |\phi_1(a)\rangle \cdots |\phi_{n-1}(a)\rangle$$

with

$$|\phi_k(a)\rangle = (|0\rangle + e^{i2\pi(0.a_k a_{k-1}\cdots a_0)}|1\rangle)/\sqrt{2},$$

where $(0.a_k a_{k-1} \cdots a_0) = a_k/2 + a_{k-1}/2^2 + \cdots + a_0/2^{k+1}$.

A quantum circuit implementing QFT usually changes the state of the kth qubit step by step in the following way.[30] Here, $R_k = \text{diag}(1, e^{i2\pi/2^k})$.

$$|a_k\rangle \xrightarrow{H} (|0\rangle + e^{i2\pi(0.a_k)}|1\rangle)/\sqrt{2}$$
$$\xrightarrow{R_2 \text{ conditioned on } a_{k-1}} (|0\rangle + e^{i2\pi(0.a_k a_{k-1})}|1\rangle)/\sqrt{2} \qquad (3)$$
$$\longrightarrow \cdots \xrightarrow{R_{k+1} \text{ conditioned on } a_0} (|0\rangle + e^{i2\pi(0.a_k a_{k-1}\cdots a_0)}|1\rangle)/\sqrt{2}.$$

Thus, the initial state $|a_k\rangle$ of each qubit evolves to $|\phi_k(a)\rangle$ without introducing entanglement with other qubits. When a superposition $|\psi_{\text{in}}\rangle = \sum_a c_a |a\rangle$ of computational basis states $|a\rangle$ is input to QFT, each qubit of each $|a\rangle$

62

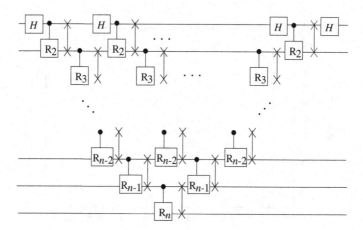

Fig. 5. Illustration of the QFT circuit introduced by Fowler *et al.*[31] for the LNN architecture.

evolves in the above manner. Therefore, the Schmidt rank for any splitting between two consecutive qubits does not exceed the degree of superposition of the input state, namely, the number of nonzero c_a's, throughout the QFT process.

The circuit structure of QFT employed in the ZKCM_QC library is the one introduced by Fowler *et al.*[31] for the linear nearest neighbor (LNN) architecture as illustrated in Fig. 5. Quantum circuits designed for the LNN architecture are suitable for TDMPS simulations because the MPS data structure for TDMPS is a sort of LNN coupling structures. One can follow how each qubit evolves by tracing the SWAP gates in the figure; it is easily verified to be in the same manner as (3).

Now we revisit an in-place addition to add an integer $b = b_{n-1}b_{n-2}\cdots b_0$ to $|a\rangle$. Note that, if b is just a classical data, it is not necessary to keep it in a classical or quantum register. The process[30] of in-place addition is as follows.

(i) $|a\rangle \overset{\text{QFT}}{\mapsto} \bigotimes_k |\phi_k(a)\rangle$.

(ii) Each qubit goes through the process:

$$|\phi_k(a)\rangle \xrightarrow{\;b_k-\text{controlled } R_1\;} (|0\rangle + e^{i2\pi(0.a_k a_{k-1}\cdots a_0 + 0.b_k)}|1\rangle)/\sqrt{2}$$

$$\xrightarrow{\;b_{k-1}-\text{controlled } R_2\;} (|0\rangle + e^{i2\pi(0.a_k a_{k-1}\cdots a_0 + 0.b_k b_{k-1})}|1\rangle)/\sqrt{2}$$

$$\longrightarrow \cdots \xrightarrow{\;b_0-\text{controlled } R_{k+1}\;} (|0\rangle + e^{i2\pi(0.a_k a_{k-1}\cdots a_0 + 0.b_k b_{k-1}\cdots b_0)}|1\rangle)/\sqrt{2}$$

$$= |\phi_k(a + b \mod 2^n)\rangle.$$

(iii) $\bigotimes_k |\phi_k(a + b \mod 2^n)\rangle \overset{\text{QFT}^{-1}}{\mapsto} |a + b \mod 2^n\rangle.$

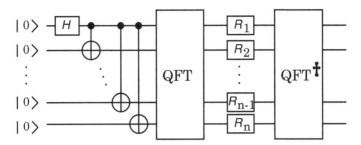

Fig. 6.   Illustration of a simple example of QFT-based addition (see the text for more explanation).

Step (ii) is a Fourier-domain addition, which alone is often used as a main component for constructing arithmetic quantum circuits.

It is now clear that the Schmidt rank of any nearest-neighbour splitting does not exceed the initial degree of superposition in the computational basis throughout the in-place addition. Now we are going to see a numerical result visualising this property.

Let us consider a simple example where $|0 \cdots 01\rangle$ is added to the GHZ state $(|0 \cdots 0\rangle + |1 \cdots 1\rangle)/\sqrt{2}$ (thus the resultant state is $(|0 \cdots 01\rangle + |0 \cdots 00\rangle)/\sqrt{2}$). The quantum circuit is illustrated in Fig. 6. A TDMPS simulation of this circuit is performed for input sizes n up to 100. As shown in Fig. 7, the maximum Schmidt rank $m_{\max,\max}$ was 2 for all $n \geq 2$; this is in accordance with the theoretical observation. The figure also shows time consumption as a function of n. It seems to be bounded from above by cn^2 with $c = 0.025$. This is reasonable as the number of quantum gates in the circuit is on the order of n^2 and the maximum Schmidt rank is constant. It should be noted that theoretical analyses of more realistic QFT-based arithmetic circuits are not so easy as the present case; there should be ancillary qubits for conditioning arithmetics and concatenations of Fourier-domain operation units. It is hoped that a theoretical estimation of the maximum Schmidt rank will be made for each QFT-based operation unit. This will be useful to analyze the QFT-based variants[31,32] of Shor's quantum prime factorization.

4.5. Simulation of the Shor's algorithm

At last, we will see the performance of the ZKCM_QC library for simulating Shor's algorithm.[23] As is well-known, prime factorization by Shor's algorithm is the most significant application of quantum computing. Known

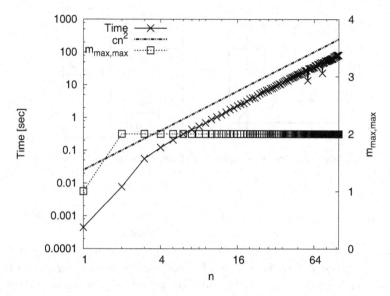

Fig. 7. Plots of running time and $m_{\mathrm{max,max}}$ in the TDMPS simulation of the quantum circuit illustrated in Fig. 6 as functions of n. The floating-point precision was set to 256. The line of cn^2 is shown with $c = 0.025$ (using the left vertical axis). The program was run as a single thread on the machine with four Intel Xeon E7-8837 2.67GHz CPUs, 315GB memory, and the RedHat Enterprise Linux 6 64-bit OS.

classical algorithms take subexponential time while Shor's algorithm takes only polynomial time for factoring a composite number.

Here, I consider Beauregard's circuit construction[33] where a semiclassical QFT is utilized (Fig. 8) to reduce the upper-side resister to a single qubit. As for the modular exponentiation, I used Fowler et al.'s construction[31] for the linear nearest neighbor (LNN) architecture, which uses LNN QFT-based Fourier-domain additions internally. In total, the circuit has $2n + 4$ qubits and $O(n^4)$ quantum gates[31] for an n-bit-long composite number N. It should be mentioned that several authors[34–36] discussed the simulability of QFT-based variants of Shor's algorithm in relation with the TDMPS and related methods, which is still an open question.

Here, several randomly-generated composite numbers have been tried and solved correctly by the TDMPS simulation of quantum prime factorization. At most a 25-bit number has been tried [thus the circuit width (the number of qubits) has been up to 54]. For this simulation, ZKCM_QC ver.0.1.2 has been used together with ZKCM ver.0.3.6. The running time looks growing only polynomially in n as shown in Fig. 9. Note that this run-

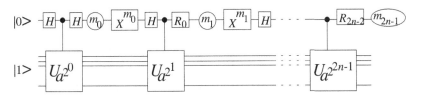

Fig. 8. Semiclassical-QFT-based quantum circuit for Shor's prime factorization introduced by Beauregard.[33]

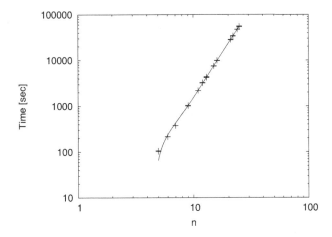

Fig. 9. Running time as a function of the bit length n of a composite number. The curve is the polynomial curve fitting with degree four. The floating-point precision was set to 128. The program was run as a single thread on the machine with four Intel Xeon E7-8837 2.67GHz CPUs, 315GB memory, and the RedHat Enterprise Linux 6 64-bit OS.

ning time is for the entire prime factorization process including the TDMPS simulation of the quantum circuit and the subsequent classical integer computation. By the fitting, it looks that $O(n^4)$ time is enough for simulation. Of course, larger n should be tried to see the tendency for a practical range, say, $n = 128$ or 256. This will be reachable by massive parallel computation in near future.

5. Concluding remarks

Several simulation results of the TDMPS simulation of quantum circuits have been introduced in this report. It has been often the case that the TDMPS method has been rather economical and able to simulate a relatively large quantum circuit within practical time. It has been indicated that the method is especially economical in simulating the quantum Fourier transform (QFT). It is interesting to further investigate the ability to simulate a QFT-based variant of quantum prime factorization by continuing the simulation work shown in Sec. 4.5. The TDMPS method, however, is not always very fast. The time consumption for a certain small 3SAT instance was significantly large in the TDMPS simulation of a single query of the Grover search. This suggests that the TDMPS simulation is not suitable for solving NP-hard database search problems although one cannot state it definitely with this example alone.

As I showed in Refs. 9 and 18, the TDMPS method is sensitive to rounding errors and truncation errors when it is used for simulating quantum computing. This is why multiprecision computation has been employed in my library. Nonetheless, owing to the poor hardware support for multiprecision floating-point operations in commercial CPUs, the constant factor of computational cost is presently significantly large. It is hoped that an automated parallelization will be implemented in the library so that it can handle larger quantum circuits feasible for solving practical computational problems.

Acknowledgments

This work was supported by the Grant-in-Aid for Scientific Research from JSPS (Grant No. 25871052).

References

1. J. Niwa, K. Matsumoto and H. Imai, *Phys. Rev. A* **66**, 062317 (2002).
2. G. Viamontes, I. Markov and J. Hayes, *Quantum Inf. Process.* **2**, 347 (2003).
3. G. Vidal, *Phys. Rev. Lett.* **91**, 147902 (2003).
4. S. Aaronson and D. Gottesman, *Phys. Rev. A* **70**, 052328 (2004).
5. A. SaiToh and M. Kitagawa, *Phys. Rev. A* **73**, 062332 (2006).
6. K. D. Raedt, K. Michielsen, H. D. Raedt, B. Trieu, G. Arnold, M. Richter, T. Lippert, H. Watanabe and N. Ito, *Comput. Phys. Comm.* **176**, 121 (2007).
7. F. Tabakin and B. Juliá-Díaz, *Comput. Phys. Comm.* **180**, 948 (2009).
8. E. Gutiérrez, S. Romero, M. Trenas and E. Zapata, *Comput. Phys. Comm.* **181**, 283 (2010).
9. A. SaiToh, *Comput. Phys. Comm.* **184**, 2005 (2013), arXiv:1303.6034.
10. J. Gruska, *Quantum Computing* (McGraw-Hill, Berkshire, UK, 1999).
11. K. A. Hallberg, *Adv. Phys.* **55**, 477 (2006).
12. U. Schollwöck, *Ann. Phys.* **326**, 96 (2011).
13. A. Kawaguchi, K. Shimizu, Y. Tokura and N. Imoto, Classical simulation of quantum algorithms using the tensor product representation *Preprint* arXiv:quant-ph/0411205.

14. C. Chamon and E. R. Mucciolo, *Phys. Rev. Lett.* **109**, 030503 (2012).

15. M. C. Bañuls, R. Orús, J. I. Latorre, A. Pérez and P. Ruiz-Femenía, *Phys. Rev. A* **73**, 022344 (2006).

16. M. R. Garey and D. S. Johnson, *Computers and Intractability: A Guide to the Theory of NP-Completeness* (W.H. Freeman and Co., New York, 1979).

17. B. Bauer *et al.*, *J. Stat. Mech.* **2011(05)**, P05001 (2011), http://alps.comp-phys.org/.

18. A. SaiToh, A multiprecision C++ library for matrix-product-state simulation of quantum computing: Evaluation of numerical errors, in *Proceedings of the Conference on Computational Physics 2012, J. Phys.: Conf. Ser.*, (IOP, London, 2013, Kobe, Japan, 14-18 October 2012). arXiv:1211.4086.

19. The GNU Multiple Precision Arithmetic Library http://gmplib.org/.

20. L. Fousse, G. Hanrot, V. Lefèvre, P. Pélissier and P. Zimmermann, *ACM Trans. Math. Software* **33**, 13 (2007), http://www.mpfr.org/.

21. D. Deutsch and R. Jozsa, *Proc. Royal Soc. London A* **439**, 553 (1992).

22. L. K. Grover, A fast quantum mechanical algorithm for database search, in *Proceedings of the 28th Annual ACM Symposium on Theory of Computing (STOC 1996)*, (ACM Press, New York, 1996, Philadelphia, PA, 22-24 May 1996).

23. P. W. Shor, *SIAM J. Comput.* **26**, 1484 (1997).

24. M. P. Zwolak, Dynamics and simulation of open quantum systems PhD Thesis, California Institute of Technology, 2008.

25. D. H. Bailey, R. Barrio and J. M. Borwein, *Appl. Math. Comput.* **218**, 10106 (2012).

26. The PARI Group, PARI/GP http://pari.math.u-bordeaux.fr/.

27. D. Mueller, *Numer. Math.* **8**, 72 (1966).

28. W. Press, S. Teukolsky, W. Vetterling and B. Flannery, *Numerical Recipes: The Art of Scientific Computing (3rd ed.)* (Cambridge University Press, Cambridge, UK, 2007).

29. M. Nakata, The MPACK (MBLAS/MLAPACK); a multiple precision arithmetic version of blas and lapack http://mplapack.sourceforge.net/.

30. T. G. Draper, Addition on a quantum computer *Preprint* arXiv:quant-ph/0008033.

31. A. G. Fowler, S. J. Devitt and L. C. L. Hollenberg, *Quantum Inf. Comput.* **4**, 237 (2004).

32. Y. Takahashi, N. Kunihiro and K. Ohta, *Quantum Inf. Comput.* **7**, 383 (2007).

33. S. Beauregard, *Quantum Inf. Comput.* **3**, 175 (2003).

34. A. Kawaguchi, K. Shimizu, Y. Tokura and N. Imoto, Classical simulation of the modular exponentiation using the tensor product decomposition in: Extended Abstract Booklet of the 11th Quantum Information Technology Symposium, Kyoto, Japan, 6-7 December 2004 (unpublished, in Japanese with English abstract), pp.163-166, technical report No. QIT2004-82.

35. D. E. Browne, *New J. Phys.* **9**, 146 (2007).

36. N. Yoran and A. J. Short, *Phys. Rev. A* **76**, 060302(R) (2007).

ENTANGLEMENT PROPERTIES OF A QUANTUM LATTICE-GAS MODEL ON SQUARE AND TRIANGULAR LADDERS

SHU TANAKA

Department of Chemistry, University of Tokyo,
7-3-1, Hongo, Bunkyo-ku, Tokyo, 113-0033, Japan
E-mail: shu-t@chem.s.u-tokyo.ac.jp
http://www.shutanaka.com/

RYO TAMURA

International Center for Young Scientists, National Institute for Materials Science,
1-2-1, Sengen, Tsukuba-shi, Ibaraki, 305-0047, Japan
E-mail: tamura.ryo@nims.go.jp
http://www.nims.go.jp/icys/ryo_tamura/tamura_home_e.html

HOSHO KATSURA

Department of Physics, Gakushuin University,
1-5-1, Mejiro, Toshima-ku, Tokyo, 171-8588, Japan
E-mail: hosho.katsura@gakushuin.ac.jp
http://www-cc.gakushuin.ac.jp/%7e20100072/

In this paper, we review the entanglement properties of a quantum lattice-gas model according to our previous paper [S. Tanaka, R. Tamura, and H. Katsura, *Phys. Rev. A* **86**, 032326 (2012)]. The ground state of the model under consideration can be exactly obtained and expressed by the Rokhsar-Kivelson type quantum superposition. The reduced density matrices of the model on square and triangular ladders are related to the transfer matrices of the classical hard-square and hard-hexagon models, respectively. In our previous paper, we investigated the entanglement properties including the entanglement entropy, the entanglement spectrum, and the nested entanglement entropy. We found that the entanglement spectra are critical when parameters are chosen so that the corresponding classical model is critical. In order to further investigate the entanglement properties, we also considered the nested entanglement entropy. As a result, the entanglement properties of the model on square and triangular ladders are described by the critical phenomena of the Ising model and the three-state ferromagnetic Potts model in two dimension, respectively.

Keywords: Entanglement; Entanglement spectrum; Nested entanglement entropy; Quantum lattice-gas model; Central charge

1. Introduction

Entanglement is a significant concept in a wide area of physics including the foundation of quantum physics, statistical physics, condensed matter physics, and computation physics.[1-3] Entanglement is a resource for quantum information processing and communication, and then, in quantum information science, it is an important issue to investigate the entanglement properties in quantum systems.[4-7] In statistical physics and condensed matter physics, theoretical models have been characterized by the entanglement properties, which will be explained later. In addition, recently, computation methods in which the concept of entanglement is used with ingenuity have been developed.[8-10]

To quantify the degree of entanglement, the von Neumann entanglement entropy and the family of Rényi entanglement entropies are often used. Let us show the relation between the von Neumann entanglement entropy and the properties of one-dimensional quantum systems. We consider a one-dimensional system whose normalized ground state is represented by $|\Psi_{g.s.}\rangle$. Here the density matrix of the ground state is given by $\rho = |\Psi_{g.s.}\rangle \langle \Psi_{g.s.}|$. We divide the whole system into two subsystems A and B. The reduced density matrix for the subsystem A is defined by

$$\rho_A := \mathrm{Tr}_B \, \rho. \tag{1}$$

The von Neumann entanglement entropy[a] is obtained by

$$\mathcal{S}_A := -\mathrm{Tr} \, \rho_A \ln \rho_A. \tag{2}$$

When the system is critical, the entanglement entropy diverges logarithmically with the number of sites in the subsystem A. The coefficient of the diverging function is decided by the central charge.[11-14] On the other hand, when the system is not critical, the entanglement entropy converges to a certain value which is equal to or less than the logarithm of the number of edge states.[12] For example, the entanglement entropy of the one-dimensional integer spin-S valence-bond-solid (VBS) state, which is the ground state of the Affleck-Kennedy-Lieb-Tasaki (AKLT) model,[15-17] converges to $2\ln(S + 1)$.[18] On top of that, the entanglement entropy of the VBS states on square and hexagonal lattices has been investigated. The entanglement entropy converges to the value less than the logarithm of the number of edge states.[19] The entanglement entropy in the thermodynamic

---

[a]Hereafter we call it entanglement entropy for simplicity.

limit for square lattice is less than that for hexoagnal lattice, since the correlation length in the VBS state on square lattice is shorter than that on hexagonal lattice.

To study the entanglement properties in more depth, the entanglement spectrum was proposed in Ref. 20. The entanglement spectrum is the eigenvalue spectrum of the reduced density matrix ρ_A. The entanglement spectrum provides more detail information about the ground state. In general, the reduced density matrix can be written by

$$\rho_A = \exp(-\mathcal{H}_E), \tag{3}$$

where $\mathcal{H}_E$ is referred to as the entanglement Hamiltonian. The above equation is similar with the relation between Hamiltonian and the density matrix in the equilibrium state at the temperature $k_B T = 1$. Entanglement spectra have been studied in many theoretical models such as quantum spin systems, quantum Hall systems, and topological insulators.[21-40] The entanglement spectrum shows an interesting behavior in some cases. For example, the relation between the entanglement spectrum of two-dimensional VBS states and the energy spectrum of other quantum system in one dimension was investigated.[39,40] Let us consider the VBS states on square and hexagonal lattices with the periodic boundary condition in one direction and the open boundary condition in the other direction. We divide the whole system into two symmetric subsystems. The boundary is parallel to the axis imposing the periodic boundary condition. The entanglement spectra of the VBS states on square and hexagonal lattices are similar with the energy dispersion of the one-dimensional antiferromagnetic and ferromagnetic Heisenberg models, respectively. For square lattice, the des Cloizeaux-Pearson mode[41] is observed in the low-lying entanglement spectrum whereas the spin-wave mode is observed in the low-lying spectrum for hexagonal lattice. The entanglement spectrum of the gapped spin-1/2 Heisenberg model on two-leg ladder looks like the energy dispersion of a single spin-1/2 Heisenberg model in one dimension.[34-36] In these cases, we can observe a beautiful fact that the reduced density matrix ρ_A can be regarded as the thermal density matrix of holographic system on one-dimensional chain, which is a significant progress in quantum statistical physics.

To further consider the entanglement properties of systems where the entanglement spectrum is similar with the energy dispersion of a holographic one-dimensional *critical* system, the nested entanglement entropy was proposed in Ref. 40. The nested entanglement entropy can determine the central charge of the critical entanglement Hamiltonian. In Ref. 40, the authors considered the nested entanglement entropy of the VBS state on square lattice. Size dependence of nested entanglement entropy indicates the central charge of the entanglement Hamiltonian for square lattice is $c = 1$ which is the central charge of the one-dimensional antiferromagnetic Heisenberg model.[40,42] This is consistent with the fact that the entanglement spectrum of the VBS state on square lattice is similar with the energy dispersion of the one-dimensional antiferromagnetic Heisenberg model.

In this paper, we review the entanglement properties of a quantum hard-core lattice gas model according to Ref. 43. We focus on the case that the ground state can be exactly obtained and is expressed by the Rokhsar-Kivelson type quantum superposition which is in the category of tensor-network states.[44–49] We consider the model which is a straightforward generalization of the model constructed in Ref. 50.

The organization of the paper is as follows. In section 2, the ground-state properties of the model are explained. In section 3, the reduced density matrix is derived. In section 4, the entanglement properties are shown. Section 5 is devoted to the summary of the paper and future perspective.

2. Model

One of the purposes of quantum information processing is to obtain the solution of problems which are hard to consider using classical computer, such as factorization problem.[51] Another important purpose of quantum information processing is to simulate strongly correlated systems such as quantum spin systems, which has been mainly attracted attention in condensed matter physics. Experimental simulation methods using cold atoms have been studied over the years.[52,53] In most cases, cold atoms in the ground state have been used for quantum control. Recently, however, quantum control using atoms excited to the Rydberg states has been considered theoretically and demonstrated experimentally. The Rydberg states offer long-ranged and strong interaction with long lifetime.[54] Because of the circumstances, it is expected that the Rydberg states can be used for implementation of quantum information processing.[55,56]

In this paper, we consider a quantum lattice-gas model with the Rydberg lattice gas in mind. The model is a straightforward generalized version of the model proposed in Ref. 50 and expresses an interacting hard-core bosonic system. Each site is occupied by a single particle or empty. Then the Hilbert space is expressed by $\bigotimes_i |n_i\rangle$. When the site i is occupied by a single particle, $n_i = 1$, whereas when the site i is empty, $n_i = 0$. Here we consider the case that there is no more than a single particle on any pair of adjacent sites, which comes from infinitely strong repulsive interaction between adjacent sites. It is convenient to introduce the spin operator with the identification: $|\uparrow\rangle \leftrightarrow |1\rangle$ and $|\downarrow\rangle \leftrightarrow |0\rangle$. Using the spin operator, the creation operator at the site i is expressed by $\hat{\sigma}_i^+ \hat{\mathcal{P}}_{\langle i\rangle}$, where the definitions of the Pauli matrices are

$$\hat{\sigma}_i^x = \begin{pmatrix} 0 & 1 \\ 1 & 0 \end{pmatrix}, \qquad \hat{\sigma}_i^y = \begin{pmatrix} 0 & -i \\ i & 0 \end{pmatrix}, \qquad \hat{\sigma}_i^z = \begin{pmatrix} 1 & 0 \\ 0 & -1 \end{pmatrix}, \qquad (4)$$

$$\hat{\sigma}_i^{\pm} = \frac{1}{2}\left(\hat{\sigma}_i^x \pm i\hat{\sigma}_i^y\right). \qquad (5)$$

The projection operator $\hat{\mathcal{P}}_{\langle i\rangle}$ represents that there is no more than one boson on any pair of adjacent sites, $i.e.$, nearest-neighbor exclusion. More precisely, the projection operator is given by

$$\hat{\mathcal{P}}_{\langle i\rangle} = \prod_{j \in G_i} \left(1 - \hat{n}_j\right), \qquad (6)$$

where $\hat{n}_j$ is the number operator at the site j, which is defined by

$$\hat{n}_j = \hat{\sigma}_j^+ \hat{\sigma}_j^- = \frac{1}{2}\left(\hat{\sigma}_j^z + 1\right), \qquad (7)$$

and G_i is a set of sites adjacent to the site i.

We consider the following Hamiltonian on a lattice Λ:

$$\hat{\mathcal{H}} = \sum_{i \in \Lambda} \hat{h}_i^\dagger(z)\hat{h}_i(z), \qquad \hat{h}_i(z) = \left[\hat{\sigma}_i^- - \sqrt{z}(1 - \hat{n}_i)\right]\hat{\mathcal{P}}_{\langle i\rangle}. \qquad (8)$$

Here we assume that z is real and nonnegative. We can rewrite the Hamiltonian:

$$\hat{\mathcal{H}} = -\sqrt{z}\sum_{i \in \Lambda}\left(\hat{\sigma}_i^+ + \hat{\sigma}_i^-\right)\hat{\mathcal{P}}_{\langle i\rangle} + \sum_{i \in \Lambda}\left[(1 - z)\,\hat{n}_i + z\right]\hat{\mathcal{P}}_{\langle i\rangle}. \qquad (9)$$

The first term in the Hamiltonian (Eq. (9)) expresses the creation and annihilation of hard-core bosons. The second term in the Hamiltonian (Eq. (9)) represents the chemical potential of the hard-core boson and interactions between hard-core bosons. The Hamiltonians given by Eq. (9) in one dimension and that on square lattice were considered in Refs. 50,57. Note that the Hamiltonian in one dimension is the transverse Ising model with next-nearest neighbor interactions and nearest-neighbor exclusion which comes from infinitely strong repulsive interaction between nearest-neighbor sites.

Next we show that the ground state of the Hamiltonian given by Eq. (9) can be exactly obtained. We consider the Hamiltonian on a general lattice Λ. The Hamiltonian is represented by the sum of local Hamiltonian $\hat{h}_i^\dagger(z)\hat{h}_i(z)$ which is positive semidefinite. Then, the Hamiltonian is also positive semidefinite and thus the eigenenergies are nonnegative. If there is a zero-energy state, the zero-energy state should be the ground state. As will be proved later, the zero-energy state is described by

$$|z\rangle = \frac{1}{\sqrt{\Xi(z)}} \prod_{i \in \Lambda} \exp\left(\sqrt{z}\hat{\sigma}_i^+ \hat{\mathcal{P}}_{\langle i \rangle}\right) |\text{vac}\rangle, \tag{10}$$

where $\Xi(z)$ is the normalization constant and $|\text{vac}\rangle$ denotes the vacuum state which is the all-down state $|\downarrow\downarrow \cdots \downarrow\rangle$ in the spin language. Here the order of the product in Eq. (10) is arbitrary since the operators $\exp(\sqrt{z}\hat{\sigma}_i^+ \hat{\mathcal{P}}_{\langle i \rangle})$ commute each other, i.e.,

$$\left[\exp(\sqrt{z}\hat{\sigma}_i^+ \hat{\mathcal{P}}_{\langle i \rangle}), \exp(\sqrt{z}\hat{\sigma}_j^+ \hat{\mathcal{P}}_{\langle j \rangle})\right] = 0, \qquad \forall i, j. \tag{11}$$

When the sites i and j are not adjacent, the commutation relation is obviously satisfied. Here we show that the commutation relation is satisfied even when the sites i and j are adjacent. The relation

$$\exp(\sqrt{z}\hat{\sigma}_i^+ \hat{\mathcal{P}}_{\langle i \rangle}) \exp(\sqrt{z}\hat{\sigma}_j^+ \hat{\mathcal{P}}_{\langle j \rangle}) = 1 + \sqrt{z}\hat{\sigma}_i^+ \hat{\mathcal{P}}_{\langle i \rangle} + \sqrt{z}\hat{\sigma}_j^+ \hat{\mathcal{P}}_{\langle j \rangle} \tag{12}$$

is derived using the following relations:

$$\exp(\sqrt{z}\hat{\sigma}_i^+ \hat{\mathcal{P}}_{\langle i \rangle}) = 1 + \sqrt{z}\hat{\sigma}_i^+ \hat{\mathcal{P}}_{\langle i \rangle}, \qquad \hat{\mathcal{P}}_{\langle j \rangle}\hat{\sigma}_i^+ = 0. \tag{13}$$

The latter relation is satisfied only when i and j are adjacent. In the same way, we can obtain

$$\exp(\sqrt{z}\hat{\sigma}_j^+ \hat{\mathcal{P}}_{\langle j \rangle}) \exp(\sqrt{z}\hat{\sigma}_i^+ \hat{\mathcal{P}}_{\langle i \rangle}) = 1 + \sqrt{z}\hat{\sigma}_j^+ \hat{\mathcal{P}}_{\langle j \rangle} + \sqrt{z}\hat{\sigma}_i^+ \hat{\mathcal{P}}_{\langle i \rangle}. \tag{14}$$

Then the commutation relation given by Eq. (11) is also satisfied even when the sites i and j are adjacent. To prove the state given by Eq. (10) is the zero-energy state, it is sufficient to show that the state $|z\rangle$ is annihilated by all $\hat{h}_i(z)$. Then,

$$\begin{aligned}
\hat{h}_i(z)|z\rangle &= \frac{1}{\sqrt{\Xi(z)}}\hat{h}_i(z) \exp\left(\sqrt{z}\hat{\sigma}_i^+ \hat{\mathcal{P}}_{\langle i \rangle}\right) \prod_{j \in \Lambda \setminus \{i\}} \exp\left(\sqrt{z}\hat{\sigma}_j^+ \hat{\mathcal{P}}_{\langle j \rangle}\right) |\text{vac}\rangle \\
&= \frac{1}{\sqrt{\Xi(z)}}\hat{\mathcal{P}}_{\langle i \rangle}\hat{\sigma}_i^- \prod_{j \in \Lambda \setminus \{i\}} \exp\left(\sqrt{z}\hat{\sigma}_j^+ \hat{\mathcal{P}}_{\langle j \rangle}\right) |\text{vac}\rangle = 0. \tag{15}
\end{aligned}$$

Here $\hat{\sigma}_i^-$ and $\exp(\sqrt{z}\hat{\sigma}_j^+ \hat{\mathcal{P}}_{\langle j \rangle})$ commute each other when sites i and j are not adjacent whereas when sites i and j are adjacent, $\hat{\sigma}_i^- \exp(\sqrt{z}\hat{\sigma}_j^+ \hat{\mathcal{P}}_{\langle j \rangle}) = \hat{\sigma}_i^-$. Then $\hat{h}_i(z)|z\rangle = 0$ since $\hat{\sigma}_i^- |\text{vac}\rangle = 0$. Since the off-diagonal elements of the Hamiltonian are nonpositive, the zero-energy state given by Eq. (10) is a unique ground state because of the Perron-Frobenius theorem.

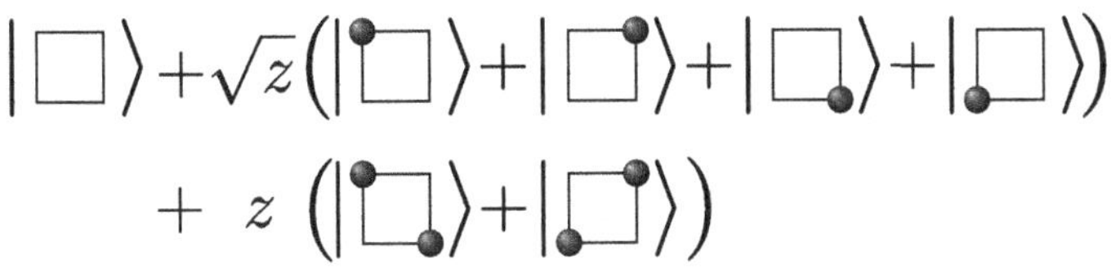

$$|\square\rangle + \sqrt{z}\left(|{\overset{\bullet}{\square}}\rangle + |\square^{\bullet}\rangle + |\square\rangle + |\underset{\bullet}{\square}\rangle\right)$$
$$+ z\left(|{\underset{\bullet}{\overset{\bullet}{\square}}}\rangle + |{\overset{\bullet}{\square}}^{\bullet}\rangle\right)$$

Fig. 1.   The unnormalized ground state of the model on a square plaquette.

To further investigate the ground-state properties, we rewrite the unnormalized ground state $|\Psi(z)\rangle = \sqrt{\Xi(z)}\,|z\rangle$ in terms of classical lattice-gas configuration as

$$|\Psi(z)\rangle = \sum_{\mathcal{C}\in\mathcal{S}} z^{n_\mathcal{C}/2}\,|\mathcal{C}\rangle, \tag{16}$$

where $|\mathcal{C}\rangle$ denote classical lattice-gas configurations of particles on the lattice Λ and $\mathcal{S}$ is the set of classical lattice-gas configurations with nearest-neighbor exclusion. $|\mathcal{C}\rangle$ are the basis states in the model and orthonormal, i.e., $\langle\mathcal{C}|\mathcal{C}'\rangle = \delta_{\mathcal{C},\mathcal{C}'}$. Furthermore, $n_\mathcal{C}$ is the number of bosons in $|\mathcal{C}\rangle$. As an example, Fig. 1 shows the ground state of the model on a square plaquette. Here the normalization constant in Eq. (10) is the partition function of the corresponding classical lattice-gas model with nearest-neighbor exclusion on the lattice Λ:

$$\Xi(z) = \langle\Psi(z)|\Psi(z)\rangle = \sum_{\mathcal{C}\in\mathcal{S}} z^{n_\mathcal{C}}. \tag{17}$$

Then, z is regarded as a fugacity of the classical lattice-gas model. The ground state is expressed by the quantum superposition of the allowed classical configurations, which is similar with the Rokhsar-Kivelson state[44] or the Valence-Bond-Solid (VBS) state[16,17] as well as their generalizations, called projected entangled pair states (PEPS).[58,59]

3. Reduced density matrix

In this paper, we consider the model on square and triangular ladders with the periodic boundary condition in the x-direction and the open boundary condition in the y-direction as shown in Fig. 2. The systems are divided into two subsystems A and B. In both cases, the boundary between two subsystems is parallel to the x-axis, which represents the dotted line in Fig. 2. The number of sites along the x-axis is L. Let $\tau = \{\tau_1, \cdots, \tau_L\}$ and $\sigma = \{\sigma_1, \cdots, \sigma_L\}$ be particle configurations in the subsystems A and B, respectively. When the site i in the subsystem A is occupied by a single

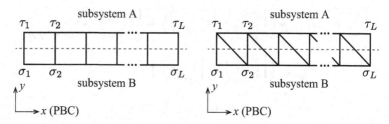

Fig. 2. (left panel) The square ladder and (right panel) the triangular ladder. The upper and lower halfs of the lattices are the subsystems A and B, respectively. The dotted lines represent the boundaries between the subsystems. The periodic boundary conditions are imposed only in the x-direction.

particle, $\tau_i = 1$, whereas the site i in the subsystem A is empty, $\tau_i = 0$. The same is applied to σ_i. The unnormalized ground state is written by

$$|\Psi(z)\rangle = \sum_\tau \sum_\sigma [T(z)]_{\tau,\sigma} |\tau\rangle \otimes |\sigma\rangle, \tag{18}$$

$$[T(z)]_{\tau,\sigma} := \prod_{i=1}^{L} w(\sigma_i, \sigma_{i+1}, \tau_{i+1}, \tau_i), \tag{19}$$

where $w(a, b, c, d)$ is the Boltzmann weight depending on the lattice structure depicted in Fig. 3. For the square ladder, $T(z)$ is represented by

$$[T(z)]_{\tau,\sigma} = \prod_{i=1}^{L} z^{(\sigma_i+\tau_i)/2}(1 - \sigma_i\tau_i)(1 - \sigma_i\sigma_{i+1})(1 - \tau_i\tau_{i+1}), \tag{20}$$

which is shown in Fig. 3 (b). For the triangular ladder, $T(z)$ is represented by

$$[T(z)]_{\tau,\sigma} = \prod_{i=1}^{L} z^{(\sigma_i+\tau_i)/2}(1 - \sigma_i\tau_i)(1 - \sigma_i\sigma_{i+1})(1 - \tau_i\tau_{i+1})(1 - \tau_i\sigma_{i+1}), \tag{21}$$

which is shown in Fig. 3 (c). Here $T(z)$ is the transfer matrix of the classical lattice-gas model on two-dimensional square and triangular lattices with nearest-neighbor exclusion.[60,61] A similar treatment was applied to the entanglement entropy of the two-dimensional Rokhsar-Kivelson wave function.[62]

To consider the entanglement properties, it is necessary to obtain the spectrum of the reduced density matrix given by

$$\rho_A = \text{Tr}_B [|z\rangle \langle z|]. \tag{22}$$

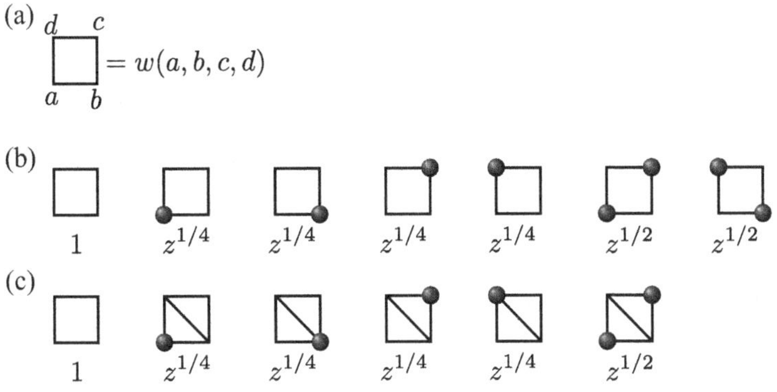

Fig. 3. (a) Graphical representation of the local Boltzmann weights. (b) Allowed states and their local Boltzmann weights for square ladder and (c) that for triangular ladder.

We adopt the approach which was applied to the two-dimensional VBS state[19,40] and considered in Ref. 63. The normalized ground state is expressed as

$$|z\rangle = \frac{1}{\sqrt{\Xi(z)}} \sum_\sigma |\phi_\sigma\rangle \otimes |\sigma\rangle, \qquad |\phi_\sigma\rangle := \sum_\tau [T(z)]_{\tau,\sigma} |\tau\rangle. \qquad (23)$$

Note that $|\phi_\sigma\rangle$ are not orthonormal. Since $|\sigma\rangle$ are orthonormal, tracing out the degrees of freedom in the subsystem B can be easily carried out:

$$\rho_A = \frac{1}{\Xi(z)} \sum_\sigma |\phi_\sigma\rangle \langle\phi_\sigma|. \qquad (24)$$

Next the Gram matrix M is introduced. The matrix elements are

$$M_{\sigma,\sigma'} = \frac{1}{\Xi(z)} \langle\phi_{\sigma'}|\phi_\sigma\rangle. \qquad (25)$$

Here $|\tau\rangle$ are also orthonormal, then,

$$M = \frac{1}{\Xi(z)} [T(z)]^T T(z), \qquad (26)$$

where the superscript T in Eq. (26) means matrix transpose. The properties of the Gram matrix M are as follows. The first one is the trace of M is unity. The second one is all the eigenvalues of M is nonnegative. The third one is all the nonzero eigenvalues of M are equal to that of ρ_A. Then, we can explore the entanglement properties by analyzing the Gram matrix instead of ρ_A.

4. Entanglement properties

We consider the entanglement properties of the model on square and triangular ladders shown in Fig. 2. The results shown in this section were obtained in Ref. 43. The parameter z-dependence of entanglement entropy, the entanglement spectrum, and the nested entanglement entropy are studied. In order to avoid the mismatch due to the boundary effect, we consider the cases that $L = 2m$ for square ladder and $L = 3m$ for triangular ladder ($m \in \mathbb{N}$).

4.1. *Entanglement entropy*

In this section, we consider the z-dependence of entanglement entropy which is obtained by

$$\mathcal{S} = -\mathrm{Tr}\,[M \ln M] = -\sum_{\alpha} p_\alpha \ln p_\alpha, \tag{27}$$

where p_α are the eigenvalues of the Gram matrix M. When the parameter z is zero, the ground state is the vacuum state and the reduced density matrix is given by

$$\rho_{\mathrm{A}} = |000000\cdots\rangle\langle000000\cdots|. \tag{28}$$

Then the entanglement entropy is equal to zero. On the other hand, in the limit of $z \to \infty$, the entanglement entropy depends on the lattice structure. In this limit, the ground state is represented by the superposition of the fully occupied states shown in Fig. 4. For square ladder, the reduced density matrix is expressed as

$$\rho_{\mathrm{A}} \sim \frac{1}{2}\left(|101010\cdots\rangle\langle101010\cdots| + |010101\cdots\rangle\langle010101\cdots|\right). \tag{29}$$

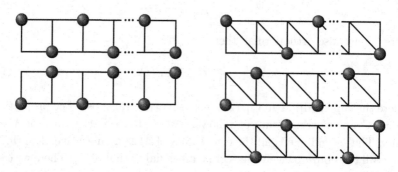

Fig. 4. The fully occupied states for square ladder (left panel) and for triangular ladder (right panel). The ground state in the limit of $z \to \infty$ is the superposition of these states with equal weights.

Then, the entanglement entropy for square ladder converges to $\ln 2$. For triangular ladder, the reduced density matrix is expressed as

$$\rho_A \sim \frac{1}{3}(|100100\cdots\rangle\langle100100\cdots| + |010010\cdots\rangle\langle010010\cdots|$$
$$+ |001001\cdots\rangle\langle001001\cdots|). \qquad (30)$$

Then, the entanglement entropy for triangular ladder converges to $\ln 3$. As shown in Figs. 4 (a) and (b) in Ref. 43, in the intermediate region between two limits, there is a peak near $z = z_c$ for finite-sized systems, where z_c denotes the critical fugacity in the corresponding classical lattice-gas model. Critical phenomena of the classical lattice-gas model with nearest-neighbor exclusion on square and triangular lattices will be reviewed in Appendix A. As explained before, the Gram matrix can be interpreted as the transfer matrix of the corresponding classical lattice-gas model. Then, it is expected that the critical entanglement properties of the original model appear at the critical point of the corresponding classical lattice-gas model. To see critical properties, we obtain the z-dependence of correlation length

$$\xi(z) = \frac{1}{\ln\left[p^{(1)}(z)/p^{(2)}(z)\right]}, \qquad (31)$$

where $p^{(1)}(z)$ and $p^{(2)}(z)$ are the largest and second largest eigenvalues of M, respectively. We find that the correlation length divided by the system size L crosses at $z = z_c$ (see Figs. 4 (e) and (f) in Ref. 43). We use the following scaling relation:

$$\xi(z)/L = f\left((z - z_c)L^{1/\nu}\right), \qquad (32)$$

where ν is the critical exponent of the correlation length. The data obtained for different system sizes collapse onto the scaling relation with $\nu = 1$ for square ladder and $\nu = 5/6$ for triangular ladder (see Fig. 5 in Ref. 43). The critical exponent of the correlation length $\nu = 1$ is the same as that of the two-dimensional Ising model whereas $\nu = 5/6$ is the same as that of the two-dimensional three-state Potts model.

From now, we focus on the entanglement properties at $z = z_c$. The entanglement entropy at $z = z_c$ linearly increases with the length L. We fit the data using the relation:

$$S(L) = \alpha L + S_0. \qquad (33)$$

As shown in Figs. 4 (c) and (d) in Ref. 43, the second term S_0 is almost zero for square and triangular ladders, which means that the topological entanglement entropy[64,65] is zero in the model.

4.2. *Entanglement spectrum*

We consider the entanglement spectrum of the model on square and triangular ladders at $z = z_c$. The entanglement spectrum was proposed in Ref. 20. As described above, the entanglement spectrum is the energy dispersion of the entanglement Hamiltonian defined by

$$\mathcal{H}_E := -\ln M. \tag{34}$$

The eigenvalues of the entanglement Hamiltonian λ_α can be calculated through the eigenvalues of the Gram matrix p_α: $\lambda_\alpha = -\ln p_\alpha$. Since we now consider the model on ladders with the periodic boundary condition in the x-direction, individual eigenstate can be labeled using the total wavenumber k because of the translational symmetry along the x-direction. The entanglement spectrum for square ladder is similar with the energy dispersion of the two-dimensional Ising model (see the top panel of Fig. 6 in Ref. 43). In addition, the entanglement spectrum for triangular ladder is similar with the two-dimensional three-state Potts model (see the bottom panel of Fig. 6 in Ref. 43). Based on the knowledge of conformal field theory (CFT), the conformal weights of low-lying states can be assigned. From the analysis, we conclude that the entanglement Hamiltonian for square ladder and that for triangular ladder at $z = z_c$ can be described by $c = 1/2$ and $c = 4/5$ CFTs, respectively. The central charge $c = 1/2$ corresponds to that of the two-dimensional Ising model whereas $c = 4/5$ does to that of the two-dimensional three-state Potts model.

4.3. *Nested entanglement entropy*

In order to confirm the fact that the entanglement Hamiltonians at $z = z_c$ can be described by the minimal CFTs with central charge $c < 1$, we consider the nested entanglement entropy proposed in Ref. 40. The nested entanglement entropy is the entanglement entropy of the ground state of the entanglement Hamilton. Let $|\phi_0\rangle$ be the ground state of $\mathcal{H}_E$. We divide the system described by $\mathcal{H}_E$ into two subsystems. In our case, the system described by $\mathcal{H}_E$ is regarded as one-dimensional quantum system. Then, the length of one of the subsystems is ℓ while on the other hand, that of the other is $L - \ell$. The nested reduced density matrix is defined by

$$\rho^{(\text{nested})}(\ell) := \text{Tr}_{\ell+1,\cdots,L}\left[|\phi_0\rangle\langle\phi_0|\right]. \tag{35}$$

The definition of the nested entanglement entropy is

$$s^{(\text{nested})}(\ell, L) := -\text{Tr}_{1,\cdots,\ell}\left[\rho^{(\text{nested})}(\ell)\ln\rho^{(\text{nested})}(\ell)\right]. \tag{36}$$

Since the entanglement spectrum at $z = z_c$ suggests that each $\mathcal{H}_E$ is critical, the nested entanglement entropy is expected to be

$$s^{(\text{nested})}(\ell, L) = \frac{c}{3} \ln [g(\ell)] + s_1, \quad g(\ell) = \frac{L}{\pi} \sin \left(\frac{\pi \ell}{L} \right), \quad (37)$$

where s_1 is a nonuniversal constant.[13] Using the relations given by Eq. (37), we obtain the central charges of $\mathcal{H}_E$ for square and triangular ladders are $c = 1/2$ and $c = 4/5$, respectively (see Fig. 7 in Ref. 43). These facts confirm that the central charge for square ladder is $c = 1/2$ and that for triangular ladder is $c = 4/5$, which was obtained from the observation of the entanglement spectrum.

5. Summary and future perspective

In this paper we reviewed the entanglement properties of a quantum hard-core lattice-gas model according to Ref. 43. The model under consideration is a generalization of the model proposed in Ref. 50. The ground state is some kind of the Rokhsar-Kivelson type state and represented by a quantum superposition. We considered the model on square and triangular ladders. The normalization constant of the ground state is the same as the partition function of the corresponding classical lattice-gas model with nearest-neighbor exclusion in two dimension. Then, we can study the entanglement properties through the analysis of the transfer matrix of the corresponding classical lattice-gas model on two-dimensional lattices. We investigated the z-dependence of entanglement entropy, where z denotes the fugacity of the classical lattice-gas model. There is a single peak near $z = z_c$, where z_c is the critial point of the corresponding classical lattice-gas model in two dimension. We also studied the entanglement spectrum and the nested entanglement entropy at $z = z_c$. The entanglement properties at $z = z_c$ for square and triangular ladders can be described by $c = 1/2$ and $c = 4/5$ CFTs, respectively.

So far, we considered the entanglement properties of the model on *ladders*. To investigate the entanglement properties of the model on *two-dimensional lattices* is a remaining problem.[66] On top of that, we focused on only the case that the ground state is exactly obtained. It is an important issue to further consider the entanglement properties for more general situations.

84

Acknowledgments

The authors would like to acknowledge Mikio Nakahara for giving them to present their studies in many workshops which were held in Kinki University. The authors met many researchers in a diverse background including physics, mathematics, and quantum information science in connection through these workshops. S.T. is also thankful for giving him an opportunity to organize this and other previous workshops, which are quite invaluable experiences. The authors are really honored that the authors celebrate the first Mikio Nakahara's KANREKI. As Mikio Nakahara said, the authors also hope that all of participants meet again in the year 2072, next Mikio Nakahara's KANREKI.

The authors thank Shunsuke Furukawa, Tohru Koma, and Tsubasa Ichikawa for valuable discussions. This work was supported by Grant-in-Aid for Japan Society for the Promotion of Science Fellows (Grant No. 23-7601), for Young Scientist (B) (Grant No. 23740298), and for Scientific Research (C) (Grant No. 25420698). R.T. is partly supported financially by National Institute for Materials Science. Numerical calculations were performed on supercomputers at the Institute for Solid State Physics, University of Tokyo.

Appendix A. Critical phenomena of the classical hard-square model

In Sec. 4.1, we considered the z-dependence of entanglement entropy and that of correlation length. We found that there is a peak in the entanglement entropy near $z = z_c$ and the correlation length divided by the number of sites along the boundary for different sizes crosses at $z = z_c$. Here z_c is the critical point of the corresponding classical lattice-gas model. In this section, we review the critical phenomena in the classical lattice-gas model. The classical lattice-gas model has been studied for a long time in statistical physics. Here we focus on the case that there is at most only single particle on any nearest-neighbor pair. The classical lattice-gas model with nearest-neighbor exclusion on square lattice is called hard-square model. The system exhibits a phase transition between a liquid-state phase at $z \ll 1$ and a solid-state phase at $z \gg 1$ as shown in Fig. A1. In the solid-state phase, a sublattice occupation is non-zero value. It was found that the second-order phase transition occurs at $z = z_c \simeq 3.796$ in Refs. 67–69. The universality class of the phase transition is the same as that of the two-dimensional Ising model. Then the correlation length critical exponent is $\nu = 1$, the

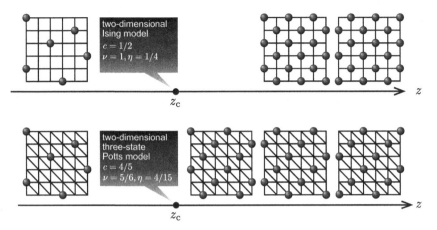

Fig. A1. A phase transition between liquid-state phase and solid-state phase in the classical lattice-gas model with nearest-neighbor exclusion.

correlation function critical exponent is $\eta = 1/4$, and the central charge is $c = 1/2$. Next we explain the classical lattice-gas model on triangular lattice called hard-hexagon model. The model is integrable and was solved in Refs. 61,68. A second-order phase transition occurs at $z = z_c = (11 + 5\sqrt{5})/2$. The universality class of the phase transition is the same as that of the two-dimensional three-state Potts model. Then, the critical exponents are $\nu = 5/6$ and $\eta = 4/15$, and the central charge is $c = 4/5$.[70]

In this way, the critical phenomena in the classical lattice-gas model with nearest-neighbor exclusion relate to the lattice structure. Depending on the lattice structure, the symmetry which breaks at the transition point is decided. In the hard-square model, twofold symmetry breaks at $z = z_c$ whereas threefold symmetry breaks at $z = z_c$ in the hard-hexagon model. Recently, in order to consider the relation between the symmetry which breaks at the transition point and nature of phase transition in a general way, a kind of lattice-gas model called Potts model with invisible states was introduced.[71–73] In the model, a first-order phase transition occurs even when the threefold symmetry breaks at the transition point in two dimension. We believe that it is an interesting topic to investigate the entanglement properties of the quantum version of the Potts model with invisible states.

References

1. L. Amico, R. Fazio, A. Osterloh, and V. Vedral, *Rev. Mod. Phys.* **80**, 517 (2008).
2. R. Horodecki, P. Horodecki, M. Horodecki, and K. Horodecki, *Rev. Mod. Phys.* **81**, 865 (2009).
3. J. Eisert, M. Cramer, and M. B. Plenio, *Rev. Mod. Phys.* **82**, 277 (2010).
4. S. Lloyd, *Science* **261**, 1569 (1993).
5. C. H. Bennet and D. P. DiVincenzo, *Nature* **404**, 247 (2000).
6. M. A. Nielsen and I. L. Chuang, *Quantum Computation and Quantum Information* (Cambridge University Press, Cambridge, 2000).
7. M. Nakahara and T. Ohmi, *Quantum Computing: From Linear Algebra to Physical Realizations* (Taylor & Francis, London, 2008).
8. G. Evenbly and G. Vidal, *Phys. Rev. Lett.* **102**, 180406 (2009).
9. G. Evenbly and G. Vidal, *Phys. Rev. Lett.* **104**, 187203 (2010).
10. K. Harada, *Phys. Rev. B* **86**, 184421 (2012).
11. C. Holzhey, F. Larsen, and F. Wilczek, *Nucl. Phys. B* **424**, 443 (1994).
12. G. Vidal, J. I. Latorre, E. Rico, and A. Kitaev, *Phys. Rev. Lett.* **90**, 227902 (2003).
13. P. Calabrese and J. Cardy, *J. Stat. Mech.* P06002 (2004).
14. P. Calabrese and J. Cardy, *Int. J. Quantum Inf.* **4**, 429 (2006).
15. I. Affleck, T. Kennedy, E. H. Lieb, and H. Tasaki, *Phys. Rev. Lett.* **59**, 799 (1987).
16. I. Affleck, T. Kennedy, E. H. Lieb, and H. Tasaki, *Commun. Math. Phys.* **115**, 477 (1988).
17. T. Kennedy, E. H. Lieb, and H. Tasaki, *J. Stat. Phys.* **53**, 383 (1988).
18. H. Katsura, M. Hirano, and Y. Hatsugai, *Phys. Rev. B* **76**, 012401 (2007).
19. H. Katsura, N. Kawashima, A. N. Kirillov, V. E. Korepin, and S. Tanaka, *J. Phys. A* **43**, 255303 (2010).
20. H. Li and F. D. M. Haldane, *Phys. Rev. Lett.* **101**, 010504 (2008).
21. N. Regnault, B. A. Bernevig, and F. D. M. Haldane, *Phys. Rev. Lett.* **103**, 016801 (2009).
22. O. S. Zozulya, M. Haque, and N. Regnault, *Phys. Rev. B* **79**, 045409 (2009).
23. A. M. Läuchli, E. J. Bergholtz, J. Suorsa, and M. Haque, *Phys. Rev. Lett.* **104**, 156404 (2010).
24. R. Thomale, A. Sterdyniak, N. Regnault, and B. A. Bernevig, *Phys. Rev. Lett.* **104**, 180502 (2010).
25. A. Chandran, M. Hermanns, N. Regnault, and B. A. Bernevig, *Phys. Rev. B* **84**, 205136 (2011).
26. X. -L. Qi, H. Katsura, and A. W. W. Ludwig, *Phys. Rev. Lett.* **108**, 196402 (2012).
27. A. M. Turner, Y. Zhang, and A. Vishwanath, *Phys. Rev. B* **82**, 241102 (2010).
28. L. Fidkowski, *Phys. Rev. Lett.* **104**, 130502 (2010).
29. E. Prodan, T. L. Hughes, and B. A. Bernevig, *Phys. Rev. Lett.* **105**, 115501 (2010).
30. P. Calabrese and A. Lefevre, *Phys. Rev. A* **78**, 032329 (2008).
31. F. Pollmann and J. E. Moore, *New J. Phys.* **12**, 025006 (2010).

32. F. Pollmann, A. M. Turner, E. Berg, and M. Oshikawa, *Phys. Rev. B* **81**, 064439 (2010).
33. R. Thomale, D. P. Arovas, and B. A. Bernevig, *Phys. Rev. Lett.* **105**, 116805 (2010).
34. D. Poilblanc, *Phys. Rev. Lett.* **105**, 077202 (2010).
35. I. Peschel and M.-C. Chung, *Europhys. Lett.* **96**, 50006 (2011).
36. A. M. Läuchli and J. Schliemann, *Phys. Rev. B* **85**, 054403 (2012).
37. H. Yao and X.-L. Qi, *Phys. Rev. Lett.* **105**, 080501 (2010).
38. C.-Y. Huang and F.-L. Lin, *Phys. Rev. B* **84**, 125110 (2011).
39. J. I. Cirac, D. Poilblanc, N. Schuch, and F. Verstraete, *Phys. Rev. B* **83**, 245134 (2011).
40. J. Lou, S. Tanaka, H. Katsura, and N. Kawashima, *Phys. Rev. B* **84**, 245128 (2011).
41. J. des Cloizeaux and J. J. Pearson, *Phys. Rev.* **128**, 2131 (1962).
42. S. Tanaka, *Interdisciplinary Information Science* **19**, 101 (2013).
43. S. Tanaka, R. Tamura, and H. Katsura, *Phys. Rev. A* **86**, 032326 (2012).
44. D. S. Rokhsar and S. A. Kivelson, *Phys. Rev. Lett.* **61**, 2376 (1988).
45. H. Niggemann, A. Klümper, and J. Zittarz, *Z. Phys. B* **104**, 103 (1997).
46. T. Nishino and K. Okunishi, *J. Phys. Soc. Jpn.* **67**, 3066 (1998).
47. Y. Hieida, K. Okunishi, and Y. Akutsu, *New J. Phys.* **1**, 7 (1999).
48. H. Niggemann, A. Klümper, and J. Zittartz, *Eur. Phys. J. B* **13**, 15 (2000).
49. M. A. Martín-Delgado, M. Roncaglia, and G. Sierra, *Phys. Rev. B* **64**, 075117 (2001).
50. I. Lesanovsky, *Phys. Rev. Lett.* **106**, 025301 (2011).
51. P. W. Shor, *SIAM J. Comput.* **26**, 1484 (1997).
52. D. Jaksch and P. Zoller, *Ann. Phys.* **315**, 52 (2005).
53. I. Bloch, J. Dailbard, and W. Zwerger, *Rev. Mod. Phys.* **80**, 885 (2008).
54. M. Saffman, T. G. Walker, and K. Mølmer, *Rev. Mod. Phys.* **82**, 2313 (2010).
55. E. Urban, T. A. Johnson, T. Henage, L. Isenhower, D. D. Yavuz, T. G. Walker, and M. Saffman, *Nat. Phys.* **5**, 110 (2009).
56. A. Gaëtan, Y. Miroshnychenko, T. Wilk, A. Chotia, M. Viteau, D. Comparat, P. Pillet, A. Browaeys, and P. Grangier, *Nat. Phys.* **5**, 115 (2009).
57. S. Ji, C. Ates, and I. Lesanovsky, *Phys. Rev. Lett.* **107**, 060406 (2011).
58. F. Verstraete and J. I. Cirac, *arXiv:cond-mat/0407066* (2004).
59. F. Verstraete and J. I. Cirac, *Phys. Rev. A* **70**, 060302 (2004).
60. R. J. Baxter, *J. Phys. A: Math. Gen.* **13**, L61 (1980).
61. R. J. Baxter, *Exactly Solved Models in Statistical Mechanics* (Academic, London, 1982).
62. J. -M. Stéphan, S. Furukawa, G. Misguich, and V. Pasquier, *Phys. Rev. B* **80**, 184421 (2009).
63. R. Jozsa and J. Schlienz, *Phys. Rev. A* **62**, 012301 (2000).
64. A. Kitaev and J. Preskill, *Phys. Rev. Lett.* **96**, 110404 (2006).
65. M. Levin and X.-G. Wen, *Phys. Rev. Lett.* **96**, 110405 (2006).
66. S. Tanaka, R. Tamura, and H. Katsura, in preparation.
67. D. S. Gaunt and M. E. Fisher, *J. Chem. Phys.* **43**, 2840 (1965).
68. R. J. Baxter, I. G. Enting, and S. K. Tsang, *J. Stat. Phys.* **22**, 465 (1980).

69. S. Todo and M. Suzuki, *Int. J. Mod. Phys.* **C7**, 811 (1996).
70. F. Y. Wu, *Rev. Mod. Phys.* **54**, 235 (1982).
71. R. Tamura, S. Tanaka, and N. Kawashima, *Prog. Theor. Phys.* **124**, 381 (2010).
72. S. Tanaka, R. Tamura, and N. Kawashima, *J. Phys.: Conf. Ser.* **297**, 012022 (2011).
73. S. Tanaka and R. Tamura, *J. Phys.: Conf. Ser.* **320**, 012025 (2011).

ON SIGNAL AMPLIFICATION FROM WEAK-VALUE AMPLIFICATION

YUTAKA SHIKANO*

*Research Center of Integrative Molecular Systems, Institute for Molecular Science,
Okazaki, Aichi 444-8585, Japan*
* E-mail: yshikano@ims.ac.jp
http://qm.ims.ac.jp

We naturally define the weak value on the basis of the probability theory in quantum mechanics. Furthermore, using this theory, we discuss the amplification process of a signal, and we have shown that there is no upper bound for the signal amplification.

Keywords: Weak measurement, Weak value, Amplification

1. Introduction

The weak value, named by Yakir Aharonov,[1] is often considered a useless physical quantity. In this paper, we show from a mathematical viewpoint that the weak value is naturally defined.

Quantum theory was introduced in 1899 with the formulation of Planck's law of black body radiation. Its formulation was initiated by Heisenberg and Schrödinger. In 1932, von Neumann mathematically formulated quantum mechanics[2] with the following postulates:

Postulate 1 (Representations of states and observables).
Any quantum system $\mathbf{S}$ *is associated with a separable Hilbert space* $\mathcal{H}_{\mathbf{S}}$, *called the state space of* $\mathbf{S}$. *Any quantum state of* $\mathbf{S}$ *is the element* ψ *of the Hilbert space and is represented in a one-to-one correspondence by a positive operator* ρ *with a unit trace, called the density operator. Any observable of* $\mathbf{S}$ *is represented in a one-to-one correspondence by a self-adjoint operator* A *densely defined on* $\mathcal{H}_{\mathbf{S}}$.

Postulate 2 (Schrödinger equation). *If* $\mathbf{S}$ *is isolated in a time interval* (t, t'), *there exists a unitary operator* U *such that if* $\mathbf{S}$ *is in* ρ *at t then* $\mathbf{S}$ *is in* $\rho = U\rho U^{\dagger}$ *at t'.*

Postulate 3 (Born formula). *Any observable A takes a value in a Borel set Δ in any ρ with the probability $Tr[E^A(\Delta)\rho]$, where $E^A(\Delta)$ is the spectral projection of A corresponding to Δ.*

Postulate 4 (Composition rule). *The composite system $\mathbf{S} + \mathbf{S}'$ is the tensor product $\mathcal{H}_{\mathbf{S}} \otimes \mathcal{H}_{\mathbf{S}'}$ of their state spaces.*

While Neumann discussed the measuring processes, he failed to obtain a mathematical postulate of measurement. Thereafter, Ozawa introduced the postulate of measurement[3] to consider the measured system and probe system.

Postulate 5 (Representation of generalized measurement). *When any observable A of a system is measured in an initial state ρ_{sys}, the state after the measurement is given by $M(\Delta)\rho_{sys} = Tr_{prob}[U(\rho_{sys} \otimes \rho_{prob})U^\dagger]$, and A takes a value in a Borel set Δ with a probability of $Tr_{sys}[\rho_{sys}M(\Delta)]$, where the time evolution operator is defined for the composite system $\mathcal{H}_{sys} \otimes \mathcal{H}_{prob}$.*

$M(\Delta)$ is often called the positive operator valued measure (POVM) or completely positive trace preserving (CPTP) map of measurement. This is a central concept of quantum information science.

In physics, measurement and state change are highly non-trivial and have been discussed by many researchers for a long time. This problem is called the measurement problem.[4] The discussion of the measurement problem results in various interpretations of quantum mechanics, especially for measuring process itself. Common examples of the various interpretations of quantum mechanics include the Copenhagen interpretation (state collapse), Bohm interpretation, and many-worlds interpretation.

Before discussing the measurement problem, we concentrate on discussing the Born rule (Postulate 3). **What is probability in quantum mechanics?** In the following section, we show the inconsistency between the classical probability theory and probability in quantum mechanics.

2. What is Probability in Quantum Mechanics?

Definition 1 (Expectation Value in Probabilistic Theory). *An expectation value $Ex[X]$ of a random variable $X(\omega)$ defined in a probabilistic space $\{\Omega, \mathcal{F}, P\}$ is*

$$Ex[X] = \int X(\omega)dP, \tag{1}$$

where dP is a probability measure.

In the conventional probability theory, the probabilistic space is $\{\Omega, \mathcal{F}, P\}$. This does not depend on the measured subject, i.e., the probability measure is independent of the observable. Of course, the random variable $X(\omega)$ is a function of Ω in $\mathbb{R}$. Conversely, in quantum mechanics, the expectation value of an observable A is given by

$$Ex[A, \psi] = \langle \psi, A\psi \rangle, \tag{2}$$

where $\psi \in \mathcal{H}$ ($||\psi|| = 1$). In order to maintain consistency with Definition 1, we consider the following definition:

Definition 2 (Expectation Value in Quantum Mechanics). *Let F be a subset of the self-adjoint operators (observables) in a Hilbert space $\mathcal{H}$ and $\psi \in \mathcal{H}$ be a state with $||\psi|| = 1$. If there exists a probabilistic space $\{\Omega, \mathcal{F}, P\}$ and a random variable $h_A(\omega)$ in $\{\Omega, \mathcal{F}, P\}$ for any observable $A \in F$ such that*

$$\langle \psi, A\psi \rangle = \int h_A(\omega) dP, \tag{3}$$

then the random variable $h_A(\omega)$ is called a functional representation of the observable $A \in F$.

We now show a simple example demonstrating the above definition. Let $\psi_t(x) \in \mathcal{H} := \mathcal{L}^2(\mathbb{R})$ be the state

$$\psi_t(x) = e^{R(t,x) + iS(t,x)}, \tag{4}$$

where $R(t, x)$ and $S(t, x)$ are real functions. Let p be the momentum operator, which is the Schrödinger operator, written as

$$p = -i\hbar\nabla, \tag{5}$$

and $F = \{x, p, p^2\}$ be a set of the observables.

Prop 2.1. The functional representation of the observable p is

$$h_p(t, x) = \hbar\nabla(R(t, x) + S(t, x)). \tag{6}$$

Prop 2.2. The functional representation of the observable p^2 is

$$h_{p^2}(t, x) = \hbar^2\left((\nabla R(t, x))^2 + (\nabla S(t, x))^2\right). \tag{7}$$

Lemma 1. *Except for the case of $\nabla R\nabla S = 0$, the relation between the functional representations of p and p^2 is*

$$(h_p)^2 \neq h_{p^2}. \tag{8}$$

In this simple example, we obtain strange results for the standard deviations of x and p. The variance of the momentum p is given by

$$Var(p) = Var(h_p) = \int \hbar^2 \left(\frac{\partial R}{\partial x} + \frac{\partial S}{\partial x} \right)^2 \mu_t dx - \left(\int \hbar \left(\frac{\partial R}{\partial x} + \frac{\partial S}{\partial x} \right) \mu_t dx \right)^2,$$
(9)

where $\mu_t = \psi_t^* \psi = e^{2R}$ is the probability density. The standard deviation of the momentum p is defined as $\tilde{\sigma}(p) := \sqrt{Var(h_p)} = \sqrt{Var(p)}$. From this definition, we obtain the following theorem.

Theorem 1 (Nagasawa Uncertainty Relationship[5]). *There does not exist a lower bound of $\tilde{\sigma}(x) \cdot \tilde{\sigma}(p)$.*

From a physical viewpoint, this result is very strange because many physicists believe that the Heisenberg uncertainty relationship,

$$\sigma(x) \cdot \sigma(p) \geq \frac{\hbar}{2},$$
(10)

is a principle of quantum mechanics. Of course, the definition of the variance differs from Eq. (9) and is

$$\sigma(A) := \sqrt{\langle \psi, \tilde{A}^2 \psi \rangle},$$
(11)

where $\tilde{A} := A - \langle \psi, A\psi \rangle I$ for any state $\psi \in \mathcal{H}$ and any observable A.

Note that the product of the standard deviations of x and p has a lower bound when $\int \nabla R \nabla S \mu_t dx = 0$, which is the exceptional case mentioned in Lemma 1. For this case, we obtain

$$\tilde{\sigma}(x) \cdot \tilde{\sigma}(p) \geq \frac{\hbar}{2}.$$
(12)

As a simple example [6, Section 53], when the state ψ is given by

$$\psi = A \exp \left(-\frac{(x - x_0)^2}{4\delta^2} + i \frac{p_0(x - x_0)}{\hbar} \right),$$
(13)

where x_0, p_0, and A are constants, the above condition $\int \nabla R \nabla S \mu_t dx = 0$ is satisfied. In addition, the inequality in Eq. (12) is satisfied. Furthermore, we obtain the following bound between the position x and the kinetic energy $p^2/2m$;

Theorem 2. *The product of the variance of the position and expectation value of the kinetic energy $p^2/2m$ has a lower bound given as*

$$\langle \psi_t, x^2 \psi_t \rangle \cdot \langle \psi_t, \frac{p^2}{2m} \psi_t \rangle \geq \frac{\hbar^2}{4} \frac{1}{2m}.$$
(14)

From a physical viewpoint, while the relation between the non-commutative observables is restricted, the position and kinetic energy are commutative. This difference originally results from the definition of random variables. Now, we answer the following question: **what is a proper expression of random variables in quantum mechanics?**

3. Extended Probability Theory via Two-State Vector Formalism

First, we discuss the strangeness of the Born rule. In conventional quantum mechanics, the expectation value of the position x is given by

$$Ex(x, \psi) := \langle \psi, x\psi \rangle = \int x\psi^*(x)\psi(x)dx = \int x|\psi(x)|^2dx, \qquad (15)$$

where $\psi(x) \in \mathcal{H} = \mathcal{L}^2(\mathbb{R})$ and $|\psi(x)|^2$ is the probability distribution and $|\psi(x)|^2dx$ is the probability measure. Furthermore, the expectation value of the momentum p is given by

$$Ex(p, \psi) := \langle \psi, p\psi \rangle = \int p\mathcal{F}[\psi^*(x)](p)\mathcal{F}[\psi(x)](p)dp = \int p|\mathcal{F}[\psi(x)](p)|^2dp, \qquad (16)$$

where $\mathcal{F}$ denotes the Fourier transform and $|\mathcal{F}[\psi]|^2dp$ denotes the probability measure, which differs from $|\psi(x)|^2dx$.

Remark 3.1 (Contextual Dependence of the Probability Measure). *The Born rule states that the probability measure depends on the observables. That is, we cannot define the probability space a priori and determine this space when determining the measured observable*[a].

In order to solve the inconsistency between physics and mathematics, we discuss the random variable. Common examples of the random variable include the results from dice and the prize price of the drawing lots. A common concept of random variables is the **accessible quantities**. In other words, in physics, the random variable must be an experimentally accessible quantity. In conventional quantum mechanics, in particular the quantum measurement theory, the experimentally quantity is real because the observable is a self-adjoint operator and only has a real spectrum. However, in quantum mechanics, we can define a complex experimentally accessible quantity, i.e., a weak value. Weak values require two states, which are

---

[a]The Kochen-Specker theorem states that the spectral representation may depend on the context, i.e., the order of taking the spectral representation of observables (See Ref.[7]). While this remark may be related to this theorem, we have not shown the direct relation.

called a pre-selected state and post-selected state, that are experimentally accessible[b].

Definition 3 (Weak Value). *The weak value of an observable A is defined as*

$$_\phi\langle A\rangle_\psi^w := \frac{\langle \phi, A\psi \rangle}{\langle \phi, \psi \rangle} \in \mathbb{C}, \tag{17}$$

where $\phi, \psi \in \mathcal{H}$ with $\int ||\phi||^2 d\phi = \int |||\psi||^2 d\psi = 1$ and $\langle \phi, \psi \rangle \neq 0$. In addition, the $\langle \phi, \psi \rangle = 0$ case is asymptotically defined[c].

The expectation value in quantum mechanics is calculated by

$$Ex(A) = \langle \psi, A\psi \rangle = \psi^* A\psi$$
$$= \int \psi^* ||\phi||^2 A\psi d\phi = \int \langle \psi, \phi \rangle \cdot \langle \phi, A\psi \rangle d\phi$$
$$= \int \frac{\langle \phi, A\psi \rangle}{\langle \phi, \psi \rangle} |\langle \phi, \psi \rangle|^2 d\phi = \int {}_\phi\langle A\rangle_\psi^w |\langle \phi, \psi \rangle|^2 d\phi, \tag{18}$$

where $h_A = {}_\phi\langle A\rangle_\phi^w$ is complex random variable and $dP = |\langle \phi, \psi \rangle|^2 d\phi$ is the probability measure that is independent of the observable A.[10] This formula indicates that the extended probability theory corresponds to the Born rule, i.e., the conventional interpretation. The variance can be obtained using

$$\mathrm{Var}(A) = \int |{}_\phi\langle A\rangle_\psi^w|^2 dP - \left(\int {}_\phi\langle A\rangle_\psi^w dP \right)^2. \tag{19}$$

In order to construct the stochastic process, we need to obtain the sequential probability, i.e., define the sequential weak value for some observables A and B as

$$_\phi\langle B, A\rangle_\psi^w := \frac{\langle \phi, BA\psi \rangle}{\langle \phi, \psi \rangle}. \tag{20}$$

Thus, we obtain the following theorem:

---

[b]The concept of weak values was proposed by Aharonov *et al.*,[8] and it has recently been developed both experimentally and theoretically. See [9, Section 1]. In order to measure the weak values, we perform the weak measurement, in which the coupling constant between the measured system and probe is very small, as well as the post-selection. Therefore, the shift in the probe observables leads to the complex weak value.
[c]Since the weak value is based on the two states, this formalism is called a two-state vector formalism. Historically speaking, the two-state vector formalism was originally motivated by the time symmetric description of quantum measurement.[11]

Theorem 3 (Sequential Process in Quantum Mechanics). *The Hilbert space $\mathcal{H}$ decomposes as $\mathcal{H} = \int \oplus_y \mathcal{H}_y$ according to the von Neumann uniqueness theorem. Let P_y be a projector from $\mathcal{H}$ to $\mathcal{H}_y$ such that $\int P_y dy = I$, where I is the identity operator on $\mathcal{H}$. Then, we can obtain the sequential process as*

$$h_{B,A} := {}_\phi \langle B, A \rangle_\psi^w = \int dy \, {}_\phi \langle P_y \rangle_\psi^w \cdot {}_\phi \langle B \rangle_y^w \cdot {}_y \langle A \rangle_\psi^w, \tag{21}$$

where $\phi, \psi \in \mathcal{H}$, and $y \in \mathcal{H}_y \subset \mathcal{H}$.

From the above consideration, the weak value can be a candidate for the fundamental physical quantities. As a result, there have been several developments on this subject.[12–21]

4. Signal Amplification

As mentioned in the previous section, the weak measurement was proposed in the context of the time-symmetric quantum measurement without collapsing the quantum state.[1] Its measurement outcome is the weak value defined in the previous section. The weak value can exceed the eigenvalue. Using this fact, the signal can be amplified. This is called the *weak-value amplification*. To study the invisible region using the standard technique, there are several studies on the weak-value amplification, as seen in Refs.[22–24] Here, the following question arises. **How can the signal be maximized?** To solve this problem, the probe wavefunction should be changed from the Gaussian distribution, which was originally used.[1] In this paper, we mathematically recapitulate the optimal probe wavefunction shown in Ref.[25] and discuss the properties of this probe wavefunction.

Let us assume the following setup: an interaction Hamiltonian coupled between the system and probe, given by

$$H_{int} = g\hat{A} \otimes \hat{p}\delta(t - t_0), \tag{22}$$

where g is the coupling constant. Here, for simplicity, we assume an instantaneous interaction at t_0. To study the maximization of the shift of the expectation value in the position space, we apply the variational principle using the Lagrangian

$$L[\xi_i(p), \xi_i^*(p), \lambda, \mu] := \langle \hat{q} \rangle_f - \lambda \left(\int dp |\xi_i(p)|^2 - 1 \right) - \mathrm{Im} \left[\mu \int dk \xi_i^*(p) \xi_i'(p) \right], \tag{23}$$

where λ and μ are Lagrange multipliers. It is noted that the gauge fixing term from the original paper[25] is added to answer the commentary paper.[26]

It is also noted that we can only find the stationary solution for the maximization of the shift of the expectation value in the position space by fixing the coupling constant g and weak value $A_w := \langle f| A |i\rangle / \langle f|i\rangle$. Furthermore, we have to set the initial expectation value of the position to zero. We have already derived the optimal probe wavefunction as the following theorem:[25]

Theorem 4. *Let the momentum space be a compact support:* $-\pi/2g \leq p \leq \pi/2g$. *In addition, the periodic boundary condition is assumed at* $p = \pm\pi/2g$ *with* $U(1)$ *degrees of freedom. Assuming that* $\hat{A}^2 = 1$ *and* $\mathrm{Re}A_w \neq 0$, *the optimal wavefunction is given by*

$$\xi_i(p) = \sqrt{\frac{g|\mathrm{Re}A_w|}{\pi}} \frac{\exp\left[-i\frac{g(|A_w|^2+1)}{2\mathrm{Re}A_w}p\right]}{\cos gp - iA_w \sin gp}, \tag{24}$$

for the case of $\mu = 0$. *Then, the shift is given by*

$$\Delta\langle\hat{q}\rangle := \langle\hat{q}\rangle_f - \langle\hat{q}\rangle_i = \langle\hat{q}\rangle_f = \frac{g(|A_w|^2+1)}{2\mathrm{Re}A_w}. \tag{25}$$

Otherwise, $\mu \neq 0$ *and there is no stationary solution.*

The proof of this theorem is given in Refs.[25,27] It is remarked that the probe wavefunction in the position space has only discrete values because the probe wavefunction is only defined in the region $-\pi/2g \leq p \leq \pi/2g$ with the periodic boundary conditions and without $U(1)$ degrees of freedom, which was used for setting the initial expectation value of the position.

The following problem will experimentally demonstrate this optimal probe wavefunction. We have not yet considered the concrete experimental setup of this situation. Specifically, for the case of light, this probe wavefunction does not satisfy the solution of the optical equation. Therefore, the wavefunction should be immediately broken after the measurement interaction because this is not a propagation mode. Within the propagation modes, e.g. Hermite-Gaussian modes and Laguerre-Gaussian modes, we have to find the optimal probe wavefunction from a practical viewpoint. For the first trial, the weak measurement in the Laguerre-Gaussian mode can be expressed as

$$\xi_{LG}(x) = C\{x + i \cdot \mathrm{sgn}(l)y\}^{|l|} \exp\left(-\frac{x^2+y^2}{4\sigma^2}\right), \tag{26}$$

where l is the azimuthal index, σ is the variance for the $l = 0$ case, and C is the normalization constant. This was considered in Refs.[28,29] However, we have not yet calculated all the higher-order effects of this mode. This is the challenging task of the weak-value amplification. As alluded to before, we need a concrete experimental setup to apply our theory. An example for such a setup is the research and development project for the gravitational wave detector. Within this, we have to find a useful case of the weak-value amplification. As far as we know, the case in which the effect is theoretically predicted but appears to be experimentally detected as the spin Hall effect of light[30] is very useful to experimentally verify the new physical theory.

Acknowledgments

I would like to thank Professor Mikio Nakahara for stimulating and valuable discussions. I would also like to convey my birthday wishes to Mikio, who is 60 years old, and I hope that our collaboration will be fruitful. Furthermore, I would like to thank Shu Tanaka, Masamitsu Bando, and Utkan Güngördü for giving me the opportunity to write this paper for publication in World Scientific Kinki University Series on Quantum Computing.

References

1. Y. Aharonov, D. Z. Albert, and L. Vaidman, Phys. Rev. Lett. **60**, 1351 (1988).
2. J. von Neumann, *Mathematische Grundlagen der Quantumechanik* (Springer, Berlin, 1932), [Eng. trans. by R. T. Beyer, *Mathematical foundations of quantum mechanics* (Princeton University Press, Princeton, 1955).]
3. M. Ozawa, J. Math. Phys. **25**, 79 (1984).
4. P. Busch, P. Mittelstaedt, and P. J. Lahti, *Quantum Theory of Measurement* (Springer-Verlag, Berlin, 1991).
5. M. Nagasawa, Kodai Math. J. **35**, 33 (2012).
6. L. Pauling and E. B. Wilson,*Introduction to Quantum Mechanics* (Mcgraw-Hill, New York, 1935).
7. C. J. Isham, *Lectures on Quantum Mechanics – Mathematical and Structural Foudantions –* (Imperial College Press, London, 1995).
8. Y. Aharonov, D. Z. Albert, and L. Vaidman, Phys. Rev. Lett. **60**, 1351 (1988).
9. Y. Shikano and A. Hosoya, J. Phys. A: Math. Theor. **43**, 025304 (2010).
10. A. Hosoya and Y. Shikano, J. Phys. A: Math. Theor. **43**, 385307 (2010).
11. Y. Aharonov, P. G. Bergmann, and J. L. Lebowitz, Phys. Rev. **134**, B1410 (1964).
12. S. Kagami, Y. Shikano, and K. Asahi, Physica E **43**, 761 (2011).
13. S. Lloyd, L. Maccone, R. Garcia-Patron, V. Giovannetti, Y. Shikano, S. Pirandola, L. A. Rozema, A. Darabi, Y. Soudagar, L. K. Shalm, and A. M. Steinberg, Phys. Rev. Lett. **106**, 040403 (2011).
14. S. Lloyd, L. Maccone, R. Garcia-Patron, V. Giovannetti, Y. Shikano, S. Pirandola, L. A. Rozema, A. Darabi, Y. Soudagar, L. K. Shalm, and A. M. Steinberg, arXiv:1108.0153.
15. S. Lloyd, L. Maccone, R. Garcia-Patron, V. Giovannetti, and Y. Shikano, Phys. Rev. D, **84**, 025007 (2011).

16. Y. Shikano, S. Kagami, S. Tanaka, and A. Hosoya, AIP Conf. Proc. **1363**, 177 (2011).

17. Y. Shikano and S. Tanaka, Europhys. Lett. **96**, 40002 (2011).

18. Y. Shikano and A. Hosoya, RIMS Kokyuroku **1658**, 257 (2009).

19. Y. Shikano and A. Hosoya, Physica E **43**, 776 (2011).

20. Y. Shikano, AIP Conf. Proc. **1327**, 482 (2011).

21. Y. Shikano, Time in Weak Value and Discrete Time Quantum Walk (LAP Lambert Academic Publishing, Germany, 2012).

22. Y. Aharonov and L. Vaidman, in *Time in Quantum Mechanics*, Vol. 1, edited by J. G. Muga, R. Sala Mayato, and I. L. Egusquiza (Springer, Berlin Heidelberg, 2008) p. 399.

23. Y. Aharonov and J. Tollaksen, in *Visions of Discovery: New Light on Physics Cosmology and Consciousness*, edited by R. Y. Chiao, M. L. Cohen, A. J. Leggett, W. D. Phillips, and C. L. Harper, Jr. (Cambridge University Press, Cambridge, 2011), p. 105.

24. Y. Shikano, in *Measurements in Quantum Mechanics*, edited by M. R. Pahlavani (InTech, 2012) p. 75, arXiv:1110.5055.

25. Y. Susa, Y. Shikano, and A. Hosoya, Phys. Rev. A **85**, 052110 (2012).

26. A. Di Lorenzo, Phys. Rev. A **87**, 046101 (2013).

27. Y. Susa, Y. Shikano, and A. Hosoya, Phys. Rev. A **87**, 046102 (2013).

28. G. Puentes, N. Hermosa, and J. P. Torres, Phys. Rev. Lett. **109**, 040401 (2012).

29. H. Kobayashi, G. Puentes, and Y. Shikano, Phys. Rev. A **86**, 053805 (2012).

30. O. Hosten and P. Kwiat, Science **319**, 787 (2008).

TOPOLOGICAL PROTECTION OF QUANTUM INFORMATION

KEISUKE FUJII

The Hakubi Center for Advanced Research, Kyoto University,
Yoshida-Ushinomiya-cho, Sakyo-ku, Kyoto 606-8302, Japan
Graduate School of Informatics, Yoshida Honmachi, Sakyo-ku, Kyoto 606-8501, Japan
* E-mail:keisuke.fujii@i.kyoto-u.ac.jp

Topologically protected fault-tolerant quantum computation on the surface code is one of the most promising routes toward the realization of scalable quantum computation. We review fault-tolerant quantum computation on the planer surface code in two dimensions and its measurement-based version, topologically protected measurement-based quantum computation in three dimensions. We also survey recent progress made based on these two models, such as topologically protected measurement-based quantum computation on the thermal states.

Keywords: topological quantum error correction, fault-tolerant quantum computation, measurement-based quantum computation

1. Introduction

The protection of quantum information from decoherence is of prime importance in the realization of quantum information processing. There have been proposed several approaches toward reliable quantum information processing, ranging from passive to active protections, such as decoherence-free subspaces,[1] dynamical decoupling,[2] and quantum error correction.[3] Among them, the most comprehensive approach is fault-tolerant quantum computation based on quantum error correction.[4] Quantum fault-tolerance theory ensures scalable quantum computation with noisy quantum devices as long as the error probability of such devices is smaller than a threshold value (see Ref. 5 and references therein). The threshold values were first obtained at almost the same time (1996) independently by Aharonov *et al.* who used a polynomial code of distance five,[6,7] by Knill *et al.* who used the Steane seven-qubit code,[8,9] by Kitaev who used toric (surface) codes.[10] All of these works are based on concatenated quantum computation, and achieve similar threshold values $\sim 10^{-6}$. Since then, fault-tolerant quantum computing has been studied as one of the most important issue in quantum information science. In 2005, Knill has proposed a novel scheme based on the C_4/C_6 error-detecting code, namely the Fibonacci scheme,[11] and achieved a considerably high threshold, a few %. Recently Fujii *et al.* have further improved this approach by making use of measurement-based quantum computation (MBQC)[12,13] on logical cluster states, which gives the highest threshold value so far, $\sim 5\%$.[14,15]

All these approaches, however, rely on availability of two-qubit gates between arbitrarily separate qubits. It was thought that if we restrict two-qubit gates to nearest-neighbor ones in two-dimension (2D), the threshold value decreases significantly.[16] The situation changed completely by the proposal made by Raussendorf *et al.*, topologically protected quantum computation on the surface code, which achieves a high threshold value $\sim 1\%$ by using only nearest-neighbor two-qubit gates in 2D.[17-19] In this paper, we review fault-tolerant quantum computation on the surface code in 2D and its measurement-based version, topologically protected MBQC in three dimensions. The we also survey recent progress made based on them.

The rest of the paper is organized as follows. In Sec. 2, we introduce the planer surface code and defect qubits on it. Specifically, we explain how to create, expand, and contract the defect region, which represents the logical qubit encoded in the planer surface code. In Sec. 3, we construct topologically protected quantum computation by braiding the defects. Topologically protected Clifford gates combined with magic state distillation allow us to achieve universal quantum computation with an arbitrary accuracy. In Sec. 4, we see how topological quantum error correction is accomplished. In Sec. 5, we translate topological quantum computation on the surface code in 2D into topologically protected MBQC in 3D, by which a beautifully simplified picture for topologically protected quantum computation is achieved. We also mention how noise is modeled and threshold values are obtained. In Sec. 6, we survey recent progress made based on topologically protected quantum computation in 2D and 3D, such as topologically protected MBQC on the thermal states. Section 7 is devoted to a concluding remark.

2. Planer surface code and defect qubits

The surface code, also called as Kitaev's toric code, was introduced by Kitaev as a toy model to understand topological order in many-body quantum systems.[20] Specifically, for later convenience, we consider the planer surface code defined on a $L \times L$ square lattice with a smooth boundary condition

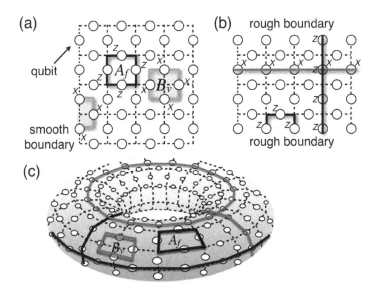

Fig. 1. (a) The planer surface code with a smooth boundary. (b) The planer surface code with a rough boundary and the logical operators, which characterize the code subspace. (c) The surface code on a torus, the so-called Kitaev's toric code. The code space has a fourfold degeneracy, which are characterized by two pairs of the logical operators consisting of products of Pauli operators on nontrivial cycles.

as shown in Fig. 1 (a), where a qubit is located on each edge. The planer surface code is defined as a stabilizer state of the face and vertex stabilizer operators,

$$A_f = \prod_{i \in E_f} Z_i, \quad B_v = \prod_{j \in E_v} X_j,$$

where Z_i and X_j denote Pauli operators on the ith and jth qubits, and E_f and E_v indicate the sets of edges surrounding a face f and adjacent to a vertex v, respectively. (Note that at the numbers of qubits included in the vertex stabilizer are three or two at the smooth boundary.) That is, the code state $|SC\rangle$ satisfies

$$|SC\rangle = A_f|SC\rangle = B_v|SC\rangle$$

for all faces f and edges v. Since the number of qubits and independent stabilizer operators are equally $2L^2 + 2L$, the planer surface code state is defined uniquely. If one choose a rough boundary or torus boundary conditions, as shown in Fig. 1 (b) and (c) respectively, we can introduce degeneracy in the surface code, which is characterized by the operators defined as products of Pauli operators on nontrivial cycles.

Here we encode quantum information into the planer surface code by introducing the defect pairs, which is developed by Raussendorf *et al.*[17-19] This allows topologically protected quantum computation by braiding the defects as seen below.

Defect qubits: In order to introduce degeneracy into the code subspace, we remove a pair of neighboring face stabilizer operators A_f and $A_{f'}$ from the stabilizer group, but their product $A_f A_{f'}$ still belongs to the stabilizer group (see Fig. 2 (a)). The degeneracy is characterized by the removed face operator A_f (or equivalently $A_{f'}$) and the operator X_i in-between two defects, which anticommute with each other and commute with all stabilizer operators. The computational bases of the logical qubit are defined as eigenstates of these operators $L_Z^p \equiv A_f$ and $L_X^p \equiv X_i$:

$$L_Z^p|\mathrm{SC}(0)\rangle = |\mathrm{SC}(0)\rangle, \quad L_Z^p|\mathrm{SC}(0)\rangle = -|\mathrm{SC}(0)\rangle,$$
$$L_X^p|\mathrm{SC}(+)\rangle = |\mathrm{SC}(+)\rangle, \quad L_X^p|\mathrm{SC}(-)\rangle = -|\mathrm{SC}(-)\rangle.$$

The logical qubit is called as a primal defect qubit, and the operators $\{L_Z^p, L_X^p\}$, which characterize the code space, are called the logical operators. Since the operator representation is rather economical, we do not write the state explicitly hereafter.

Similarly to the above case, we can also define a defect qubit by removing the vertex stabilizer operator, which we call a dual defect qubit. The logical operators are given by $L_X^d \equiv B_v$ and $L_Z^p \equiv Z_i$ (see Fig. 1 (a)).

Defect pair creation: In quantum computation, we have to prepare the logical qubit in the logical Pauli basis. This corresponds to creating a defect qubit on the planer surface code without any defect. In the case of a primal defect qubit, the planer surface code state itself can be regarded as the L_Z^p-basis state, since the code state is the eigenstate of L_Z^p with eigenvalue $+1$. This is also the case for the L_X^d-basis state of the dual defect qubit.

On the other hand, the L_X^p-basis state is created by measuring the qubit in the $L_X^p = X_i$-basis. Since A_f and $A_{f'}$ anticommute with $L_X^p = X_i$, they are removed from the stabilizer group, while $A_f A_{f'}$ commutes with $L_X^p = X_i$ and still belongs to the stabilizer group. If the measurement outcome is -1, by performing $L_Z^p = A_f$ (logical phase flipping), we can obtain the eigenstate of L_X^p with eigenvalue $+1$. (In many cases there is no necessity to do this. It is sufficient to record the measurement outcomes.) Similarly, we can also prepare the L_Z^d-basis state for the dual pair qubit.

In order to perform universal quantum computation, in addition to the

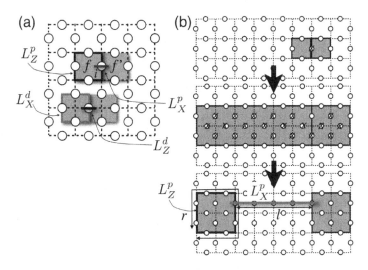

Fig. 2. (a) The primal and dual defect pair qubits and the logical operators. (b) The defect creation, expansion, and contraction from top to bottom.

Pauli-basis states, we need a non-Pauli-basis state in order to implement a non-Clifford gate. For the primal defect qubit, we first prepare the L_Z^p-basis state (i.e. the planer surface), and then perform a $e^{i\theta X_i}$ rotation. Since $L_X^p = X_i$ is the logical operator, this works as a rotation for the logical qubit. Similarly, a rotation for the dual defect qubit can be done by $e^{i\theta Z_i}$.

Defect expansion and move: The above defect qubits are not topologically protected yet, since their logical operators are given by four- or one-body Pauli operators, and hence there is no error-tolerance. To protect quantum information topologically, we have to expand the region of the defect and separate one defect from the other. In the case of the primal defect qubit, we can expand the defect region by measuring the qubits that are to be newly included into the defect region in the X-basis, since they anticommute with the face stabilizer operators. Then a new logical L_Z^p operator is defined as a product of Z operators on a cycle wrapping around the defect as shown in Fig. 2 (b). We should note that during this process, information with respect to the operator L_Z^p is not disturbed. For the contraction of the defect, we measure the face stabilizer operators on the defects that are to be contracted.

By using the above defect expansion and contraction, we can separate one defect from the other arbitrarily. Then a new logical operator L_X^p is

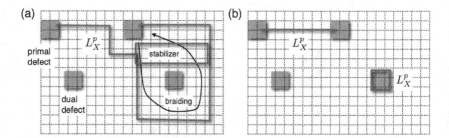

Fig. 3. (a) The braiding operation of a primal defect around a dual defect. The logical L_X^p operator wraps around the dual defect after the braiding. (b) The logical L_X^p operator after the braiding is equivalent to $L_X^p \otimes L_X^d$ before the braiding up to the multiplication of a stabilizer operator.

defined as a product of Xs connecting the two defects as shown in Fig. 2 (b). If the logical operators have the same topology, by which we mean that they are equivalent up to some multiplications of stabilizer operators, then their actions on the logical qubit are exactly the same. Note that during the defect expansion and contraction, information with respect to L_X^p is not disturbed, since all operations are commute with it. Similarly, for the dual defect qubit, we can arbitrarily expand the defect region and separate two defects. Let us define a code distance d as the minimum number of qubits in the logical operator on which Pauli operators act nontrivially. Specifically, d is given by $d = \min(l, r)$, where l and r are the distance between the defect and the perimeter of the defect region, respectively (see Fig. 2 (b)). Roughly speaking, if the number of errors is smaller than $\lfloor d/2 \rfloor$, the error correction succeeds.

3. Topological quantum computation by braiding

In order to perform universal quantum computation, we need both Clifford gates, such as controlled-Z (CZ: $\Lambda(Z)$), controlled-NOT (CNOT: $\Lambda(X)$), Phase ($S = e^{i(\pi/4)Z}$) and Hadamard gates (H), and non-Clifford gates, such as $\pi/8$ and Toffoli gates. Here we choose a universal set of gates, $\{\Lambda(X), S, \pi/8\}$, specifically.

The CNOT gate: We first consider the CNOT gate between primal (control) and dual (target) qubits. Suppose there are primal and dual qubits on the planer surface code. By using the defect expansion and contraction mentioned previously, we can move the defect toward where we want. Let us braid the primal defect around the dual defect as shown in Fig. 3 (a). The logical L_X^p operator after the braiding is wrapping around the dual defect.

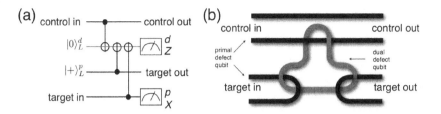

Fig. 4. (a) The teleportation-based CNOT gate between primal defect qubits. (b) The logical CNOT gate between two primal defect qubits that is equivalent to the circuit given in (a).

By multiplying a stabilizer operator as shown in Fig. 3 (a), this operator is equivalently rewritten as a product of the logical operators, $L_X^p \otimes L_X^d$. Thus the logical $L_X^p \otimes I^d$ operator before the braiding is transformed into $L_X^p \otimes L_X^d$ by the braiding operation. A similar observation on the logical L_Z^d operator tells us that the logical $I^p \otimes L_Z^d$ operator is transformed into a product of the logical operators, $L_Z^p \otimes L_Z^d$. This transformation rule is exactly the same as that of the CNOT gate, and hence we understand the braiding operation results in the logical CNOT gate for the primal (control) and dual (target) defect qubits.

Unfortunately, the CNOT gates of which primal (dual) defect qubits are always control (target) qubits, are always commutable. This is a natural consequence of fact that the defect qubits on the planer surface code are Abelian. In order to realize the CNOT gates between primal qubits (and hence noncommuting gates), we utilize a teleportation-based gate[21] as shown in Fig. 4 (a), where the CNOT gates only between primal (control) and dual (target) qubits are employed. The X-basis measurement of the primal qubit is done by a complete contraction of two defects: bringing the two defects side by side and measuring the qubit in-between the defects in the X-basis, which tells us the eigenvalue of L_X^p. The Z-basis measurement of the dual qubit is also done in a similar way. Accordingly, the CNOT gate between primal qubits is realized by braiding the primal defects around a dual defect and annihilating the dual defect pair appropriately as shown in Fig. 4 (b).

S and $\pi/8$ gate: The S $(= e^{i(\pi/4)Z})$ and $\pi/8$ $(= e^{i(\pi/8)Z})$ gates are also implemented by gate teleportation using the CNOT gate and the Pauli basis measurements, where the logical Y-basis state and logical $H = (X+Z)/\sqrt{2}$-basis state are used as ancillae in teleportation.[21,22] These ancilla states are called singular qubits, since they are not protected topologically. The

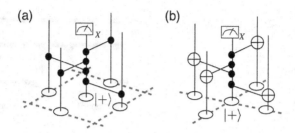

Fig. 5. (a) The syndrome measurement for the face stabilizer. (b) The syndrome measurement for the vertex stabilizer.

singular qubits are prepared by the method mentioned previously. Specifically, the Y- and H-basis states for the primal qubit are prepared by $e^{i(\pi/4)X}$ and $e^{i(\pi/8)X}$ rotations on the planer surface codes, respectively. The errors on the singular qubits cannot be removed by topological quantum error correction, since we cannot distinguish an erroneous rotation $e^{i(\theta+\epsilon)X}$ with a correct logical rotation $e^{i\theta X}$. Hence we need another special trick to protect them from errors. Fortunately, for the Y- and H-basis states, we can utilize magic state distillation to distill ideal magic states from many copies of noisy magic states using topologically protected Clifford gates.[23]

4. Topological error correction and threshold values

Error correction on the surface code is performed by using the eigenvalues of the stabilizer operators, which we call the error syndrome. An eigenvalue of a face stabilizer operator is measured by using an ancilla qubit $|+\rangle$, CZ gates, and the X-basis measurement as shown in Fig. 5 (a). Similarly a vertex stabilizer is measured by using CNOT gates as shown in Fig. 5 (b). The resultant error syndrome is analyzed to infer the location of errors.

Let us first consider the simplest case, where X and Z errors occur on each qubit with an independent and identical probability p, and there is no error during the syndrome measurements. To describe errors and their syndrome, it is convenient to use some mathematical terminology. By assigning a $\mathbf{Z}_2$ value $c_i \in \{0,1\}$ for each edge i of the square lattice, we define a $\mathbf{Z}_2$-valued one-chain c. (By abuse of notation, we also use the term one-chain c as a set of edges on which $c_i = 1$ is assigned.) The set of one-chains $\{c\}$ form a vector space over $\mathbf{Z}_2$. We also define one-chains $\{\bar{c}\}$ on the dual square lattice, where each primal face f, edge i, and vertex v are mapped into dual vertex $\bar{v}$, edge $\bar{i}$, and face $\bar{f}$, respectively. The one-chains on primal and dual square lattices are called primal and dual one-chains, respectively.

Z (X) errors are represented by a primal (dual) one-chain c^z $(\bar{c}^x)$, where a Z (X) error is located on the edge i (dual edge $\bar{i}$) with $c_i^z = 1$ $(\bar{c}_{\bar{i}}^x = 1)$. Z errors are detected at the boundary ∂c^z of the primal one-chain c^z, since $|c^z \cup \partial \bar{f}| = $ odd for the vertex $\bar{f} = v \in \partial c^z$ (that is, B_v anticommutes with such Z errors). Similarly, X errors are detected at the boundary $\partial \bar{c}^x$ of the dual one-chain $\bar{c}^x$, since $|\bar{c}^x \cup \partial f| = $ odd for the face $\bar{f} \in \partial \bar{c}^x$ (that is, A_f anticommutes with such X errors). In error correction, using the error syndrome ∂c^z and $\partial \bar{c}^x$, we estimate one-chains r^z and $\bar{r}^x$ for the recovery such that $r^z + c^z$ and $\bar{r}^x + \bar{c}^x$ result in trivial cycles (recall that trivial cycle operators act on the code space trivially, since they belong to the stabilizer group).

This inference problem is beautifully mapped into the spin glass theory.[24] More precisely, the posterior probability of the errors in the Bayes estimation and the logical error probability are respectively mapped into a partition function of the random-bond Ising model and its quenched average, where antiferromagnetic bonds correspond to the location of errors. Topological error correction succeeds in the ferromagnetic phase (see for example[25]). The optimal threshold value, which corresponds to the critical point on the Nishimori line,[26] is estimated to be 10.9%[27] theoretically, which is in good agreement with the numerical result.[28]

Since errors are detected at the boundary of the one-chains, the location of errors, that is, the one-chains r^z and r^x satisfying $\partial r^z = \partial c^z$ and $\partial r^x = \partial c^z$, can be intuitively given by a shortest path that connects elements in the boundary, i.e., faces and vertices of the eigenvalue -1. Thus the location of errors can be obtained efficiently by using by using the Edmonds' minimum-weight-perfect-matching algorithm.[29] (This corresponds to find the ground state configuration of the random-bond Ising model.) While it is not optimal, the threshold value is very close to the optimal one 10.3%.[30]

In fault-tolerant quantum computation, we have to take all sources of errors into account, such as imperfections during gate operations for the syndrome measurements. We will mention fault-tolerant thresholds taking such errors after introducing topologically protected MBQC in the next section.

5. Topologically protected MBQC

Topological quantum computation on the surface code is originally proposed as MBQC on the three-dimensional cluster state. Actually, in MBQC, we can achieve a beautifully simplified picture of topologically protected quantum computation as seen below.

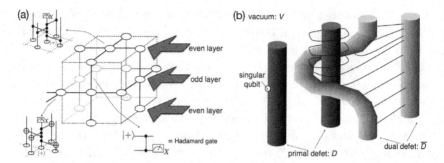

Fig. 6. (a) A unit cell of the 3D cluster state for topologically protected MBQC. Correspondences between the circuit-based 2D scheme are shown explicitly. (b) Vacuum V, primal defect D, dual defect $\bar{D}$ regions and singular qubits. A time evolution of the logical L_Z^d operator during the braiding operation is also shown.

We first explain why topological error correction on the surface code is mapped into an MBQC on the 3D cluster state, where qubits are located on the faces and edges of the cubic lattice as shown in Fig. 6 (a). As mentioned before, a face stabilizer measurement consists of an ancilla qubit $|+\rangle$ located at the face center, CZ gates between the ancilla qubit and four qubits on the face, and the X-basis measurement on the ancilla qubit (see Fig. 5 (a)). This corresponds to the cluster state located on the face of the even layer as shown in Fig. 6 (a). For the vertex stabilizer, we use an ancilla qubit $|+\rangle$ located on the vertex, CNOT gates between the ancilla qubit (control) and four qubits (target) adjacent to the vertex, and the X-basis measurement on the ancilla qubit. By using Hadamard gate, the CNOT gate is transformed into CZ gate. Thus the vertex stabilizer measurement corresponds to the cluster state located on the dual face (face of the dual cubic lattice) on the odd layer, which is transferred to the next even layer with the Hadamard transformation as shown in Fig. 6 (a).

These observations lead that the syndrome measurements for topological quantum error correction can be implemented solely by the X-basis measurements on the 3D cluster state, where the face qubits on the odd layer and edge qubits on the even layer constitute the surface code state. The face qubits on the even layer and the edge qubits on the odd layer correspond to the ancillae for the face and vertex stabilizer measurements, respectively (see Fig. 6 (a)).

Next we consider how measurements utilized in the defect creation, expansion, and contraction on the surface code are mapped in MBQC. In order to create and expand the primal defect pair, as mentioned previously, we measure the qubit in-between the defects in the X-basis. Since the cluster state on the even layer is directly related to the surface code in 2D (the cluster state on the odd layer is related under the Hadamard transformation), the primal defect creation and expansion can be done X-basis measurements of the associated qubit on the even layer, where we have to cut the time-like (vertical) cluster-bond connecting to the next layer, which can be done by the Z-basis measurement of the corresponding qubit on the next odd layer. On the other hand, dual defect creation and expansion can be done simply by the Z-basis measurement of the corresponding qubit on the even layer. The defect contraction for the primal and dual defect qubits are merely done face and vertex stabilizer measurements, and hence they are equivalent to the X-basis measurements mentioned.

Finally, we consider singular qubit preparations for the non-Clifford gate. Let us recall that the singular qubits for the primal and dual defect are prepared by $e^{i\theta X}$ and $e^{i\theta Z}$ rotations on the surface code. The singular qubit preparation for the primal qubit corresponds to the $e^{i\theta Z}$ rotation of the associated qubit on the odd layer. The following X-basis measurement appropriately transforms it into $e^{i\theta X}$ on the next even layer. On the other hand, the singular qubit preparation for the dual qubit is simply done by $e^{i\theta Z}$ of the associated qubit on the even layer.

Now we can describe topologically protected MBQC in a beautifully simplified way. The 3D cluster state is divided into disjoint regions, vacuum V, primal defect D, dual defect $\bar{D}$ regions, and singular qubits S (see Fig. 6 (b)). All qubits in the vacuum region V are measured in the X-basis. In the primal defect region D, the face qubits are measured in the Z-basis, while the edge qubits are measured in the X-basis. In the dual defect region $\bar{D}$, the edge qubits are measured in the Z-basis, while face qubits are measured in the Z-basis. Then the singular qubits are measured in the Y- or $H = (X + Z)/\sqrt{2}$-bases depending on the states which we want to prepare. This simplification was obtained in the original paper,[17] and then it was translated into the circuit-based 2D scheme.[18,19,31]

In this picture, the syndrome measurement and topological error correction are also reformulated simply. The 3D cluster state is stabilized by the following cluster stabilizers defined for each faces f and $\bar{f}$ on the primal and dual cubes, respectively:

$$K(f) = X_f \prod_{e \in \partial f} Z_e, \quad K(\bar{f}) = X_{\bar{f}} \prod_{\bar{e} \in \partial \bar{f}} Z_{\bar{e}},$$

where e, and $\bar{e}$ indicate edges on the primal and dual cubes, respectively. By multiplying the stabilizers on each cube, we obtain the check operators defined on each primal and dual cubes,

$$\prod_{f\in\partial q} K(f) = \prod_{f\in\partial q} X_f, \quad \prod_{f\in\partial q} K(\bar{f}) = \prod_{\bar{f}\in\partial\bar{q}} X_{\bar{f}}.$$

Since the 3D cluster state is stabilized by these operators, the parity of six X-basis measurement outcomes on a cube is always even. On the other hand, as mentioned above, topological error correction is done in the vacuum region V by the X-basis measurements. Thus errors on the vacuum region, which appear as Z errors, can be detected as odd parities of the check operators. The errors for the Z-basis measurements in the defect regions, which appear as X errors, do not cause any problem except for the boundary in-between the defect and vacuum regions. The errors on the boundary of the defect region effectively enhance the Z errors on the neighboring vacuum region through the Pauli byproducts in MBQC. Thus this can be also detected by the check operators. Finally, errors on the singular qubits are removed by magic state distillation executed by the topologically protected Clifford gates as mentioned before.

In topological quantum error correction, it is sufficient to estimate the location of the Z errors in the vacuum region using the parities of the check operators. Similarly to the 2D case, Z errors are detected by the check operators located at the boundary of a one-chain associated with the Z errors. Thus we can estimate the location of the errors by using the Edmonds' minimum-weight-perfect-matching algorithm,[29] which results in the threshold value 2.9% for an independent Z error. Furthermore, this inference problem can be mapped into the random-plaquette $\mathbf{Z}_2$ gauge model, where the Higgs and confinement phases correspond to fault-tolerant and non-fault-tolerant regions.[24] The optimal threshold value is estimated to be 3.3% by a numerical simulation.[32]

Fault-tolerant threshold values have been calculated assuming depolarizing noise for each elementary operations: (i) An ideal single-qubit gate is followed by single-qubit depolarizing noise,

$$(1 - p_1)\rho + \sum_{A\in\{X,Y,Z\}} \frac{p_1}{3} A\rho A.$$

(ii) An ideal two-qubit gate is followed by two-qubit depolarizing noise,

$$(1 - p_2)\rho + \sum_{A,B\in\{I,X,Y,Z\}\setminus I\otimes I} \frac{p_2}{15}(A\otimes B)\rho(A\otimes B).$$

(iii) The Pauli basis state preparations and measurements are implemented with an error probability p_P and p_M, respectively. For example, when these error probabilities are parameterized as $p_1 = p_2 = (3/2)p_P = (3/2)p_M$ as done in Refs.,[17-19] the fault-tolerant threshold value is obtained by numerical simulations to be $p_2 = 0.75\%$ and $p_2 = 0.67\%$ for the 2D and 3D cases, respectively. The slight difference between these two values comes from the fact that the Hadamard gate is implemented by teleportation in the 3D case. Roughly speaking, the total error probability per qubit in the 3D case amounts to be $p_P + p_M + 4p_2$, corresponding to the X-basis state preparation and measurement, and four CZ gates constructing the 3D cluster state. From the threshold value 2.9% for the independent Z error, we impose $p_P + p_M + 4p_2 \leq 2.9\%$ and obtain $p_2 \leq 0.54\%$, which is in good agreement with the detailed one, 0.67%.

6. Applications

Topologically protected quantum computation in 2D and 3D have been utilized as platforms to design fault-tolerant architectures for quantum computation. One promising approach is on-chip monolithic architectures, such as quantum dots[33,34] and superconducting qubits,[35,36] where huge number of qubits are integrated on a single chip and each individual qubits and interactions between them are manipulated by multiple lasers or electrodes. Another approach is a distributed modular architecture, where a-few-qubit local modules are connected with quantum channels mediating interactions between separate modules.[37-43] The a-few-qubit quantum module has been experimentally realized already in various physical systems, such as nitrogen-vacancy centers in diamond, and trapped ions. Furthermore, entangling operations between separate local modules have been experimentally demonstrated. These experimental and theoretical progress will gradually lead us to large scale quantum computation.

Topologically protected MBQC in 3D is useful to study quantum computational capacity of quantum many-body states at finite temperature.[44] Barrett $et\ al.$ have proposed the free-cluster Hamiltonian:

$$H_{\text{fc}} = -J \sum_{f,\bar{f}} [K(f) + K(\bar{f})],$$

whose ground state is exactly the same as the 3D cluster state.[45] The thermal state at temperature $T = 1/(\beta J)$ is given by

$$\rho_{\text{fc}} = e^{-\beta H_{\text{fc}}} / \text{Tr}[e^{-\beta H_{\text{fc}}}]$$
$$= U_{\mathcal{T}} e^{-\beta H_{\text{f}}} U_{\mathcal{T}}^{\dagger} / \text{Tr}[e^{-\beta H_{\text{f}}}],$$

where $H_{\mathrm{f}} \equiv -J \sum_i X_i = U_{\mathcal{T}} H_{\mathrm{fc}} U_{\mathcal{T}}^\dagger$, and $U_{\mathcal{T}} \equiv \prod_{\langle f\bar{f}\rangle} \Lambda_{f\bar{f}}(Z)$, i.e., a product of CZ gates for all cluster bonds. Since H_{f} is an interaction-free Hamiltonian, this model does not undergo any thermodynamic phase transition.

The thermal state, on the other hand, can be given as an ideal 3D cluster state followed by an independent dephasing for each qubit with probability $p = e^{-2\beta J}/(1 + e^{-2\beta J})$. Thus if $p \leq 2.9 - 3.3\%$ and hence $T = 1/(\beta J) \leq 0.57 - 0.59$, then we can perform universal quantum computation reliably in the presence of errors originated from the thermal excitation. This means that this model exhibits a transition of computational capability, while there is no thermodynamic phase transition in the physical system.

While the above Hamiltonian employs multi-body interactions, the 3D cluster stat can be generated from thermal states of nearest-neighbor two-body Hamiltonians of spin-2 or spin3/2 particles by local filtering operations.[46,47]

On the other hand, Fujii $et\ al.$ have proposed an interacting cluster Hamiltonian,[48]

$$H_{\mathrm{ic}} = -J \sum_{\langle f,\bar{f}\rangle} K(f)K(\bar{f}).$$

Since interactions between the cluster stabilizers are introduced, this model is mapped by $U_{\mathcal{T}}$ into an Ising model on a 3D lattice. Thus it undergoes a thermodynamic phase transition at a finite temperature. The degenerate ground states are again the 3D cluster state, up to the simultaneous spin flipping due to the global symmetry. Since in a ferromagnetic ordered phase the eigenvalues of the cluster stabilizer have a long range order (they are likely to be aligned in the same direction), topologically protected MBQC on the symmetry breaking thermal state has special robustness against the thermal excitations. In Ref. 48 topological error correction in this model is mapped to a correlated random plaquette $\mathbf{Z}_2$-gauge model in 3D, where disorder in the signs of the plaquettes has an Ising-type correlation. By using this properties and the gauge transformation on the Nishimori line,[26] they showed that the critical temperature of this model, and hence the threshold temperature for topological protection, is equal to the critical temperature of the 3D Ising model, which is the unitary equivalent model of the interacting cluster Hamiltonian. This means that the critical temperatures for the topological protection and the thermodynamic phase transition of the underlying physical system coincides exactly. Due to this fact, we can improve the threshold temperature for topological protection by one order of magnitude.

7. Conclusion

We have reviewed topologically protected quantum computation on the surface code in 2D and its measurement-based version. These models are quite simple only requiring logical two-qubit gates in 2D or 3D. Nevertheless the threshold values are reasonably high. This is the main reason why these models are employed as platforms for architecture designs for scalable quantum computation. There are a lot of works related to fault-tolerant quantum computation on the surface code that cannot be covered in this paper. A part of them can be found in Ref. 25.

Acknowledgments

The author thanks M. Nakahara and S. Tanaka for giving me a nice opportunity through the workshop *"Symposium on Quantum Computing, Thermodynamics, and Statistical Physics"* to collaborate with M. Ohzeki for the work in Ref. 48. This work is supported by JSPS Grant-in-Aid for Research Activity Start-up 25887034.

References

1. D. A. Lidar, I. L. Chuang and K. B. Whaley, *Phys. Rev. Lett.* **81**, 2594 (1998).
2. L. Viola, E. Knill and S. Lloyd, *Phys. Rev. Lett.* **82**, 2417 (1999).
3. P. W. Shor, *Phys. Rev. A* **52**, R2493 (1995).
4. D. P. DiVincenzo and P. W. Shor, *Phys. Rev. Lett.* **77**, 3260 (1996).
5. M. A. Nielsen and I. L. Chuang, *Quantum Computation and Quantum Information* (Cambridge University Press, 2000).
6. D. Aharonov and M. Ben-Or, Fault tolerant quantum computation with constant error, in *Proc. ACM STOC*, 1997.
7. D. Aharonov and M. Ben-Or, *SIAM J. Compt.* **38**, 1207 (2008).
8. E. Knill, R. Laflamme and W. H. Zurek, *Proc. R. Soc. London A* **454**, 365 (1998).
9. E. Knill, R. Laflamme and W. H. Zurek, *Science* **279**, 342 (1998).
10. A. Y. Kitaev, *Russ. Math. Surv.* **52**, 1191 (1997).
11. E. Knill, *Nature (London)* **434**, 39 (2005).
12. R. Raussendorf and H. J. Briegel, *Phys. Rev. Lett.* **86**, p. 5188 (2001).
13. R. Raussendorf, D. E. Browne and H. J. Briegel, *Phys. Rev. A* **68**, p. 022312 (2003).
14. K. Fujii and K. Yamamoto, *Phys. Rev. A* **81**, 042324 (2010).
15. K. Fujii and K. Yamamoto, *Phys. Rev. A* **82**, 060301 (2010).
16. K. M. Svore, B. M. Terhal and D. P. DiVincenzo, *Phys. Rev. A* **72**, 022317 (2005).
17. R. Raussendorf, J. Harrington and K. Goyal, *Ann. Phys.* **321**, 2242 (2006).
18. R. Raussendorf and J. Harrington, *Phys. Rev. Lett.* **98**, 190504 (2007).

19. R. Raussendorf, J. Harrington and K. Goyal, *New J. Phys.* **9**, 199 (2007).
20. A. Y. Kitaev, *Ann. Phys.* **303**, 2 (2003).
21. D. Gottesman and I. L. Chuang, *Nature* **402**, 390 (1999).
22. X. Zhou, D. W. Leung and I. L. Chuang, *Physical Review A* **62**, p. 052316 (2000).
23. S. Bravyi and A. Kitaev, *Physical Review A* **71**, p. 022316 (2005).
24. E. Dennis, A. Y. Kitaev, A. Landahl and J. Preskill, *J. Math. Phys.* **43**, p. 4452 (2002).
25. K. Fujii, *Interdisciplinary Information Sciences* **19**, 1 (2013).
26. H. Nishimori, *Statistical Spin Glasses and Information Processing: An introduction* (Oxford University Press, 2001).
27. M. Ohzeki, *Physical Review E* **79**, p. 021129 (2009).
28. F. Merz and J. Chalker, *Physical Review B* **65**, p. 054425 (2002).
29. J. Edmonds, *Canadian Journal of mathematics* **17**, 449 (1965).
30. C. Wang, J. Harrington and J. Preskill, *Annals of Physics* **303**, 31 (2003).
31. A. G. Fowler, A. M. Stephens and P. Groszkowski, *Physical Review A* **80**, p. 052312 (2009).
32. T. Ohno, G. Arakawa, I. Ichinose and T. Matsui, *Nuclear Physics B* **697**, 462 (2004).
33. R. VanMeter, T. D. Ladd, A. G. Fowler and Y. Yamamoto, *Int. J. Quantum Inform.* **8**, 295 (2010).
34. N. C. Jones, R. Van Meter, A. G. Fowler, P. L. McMahon, J. Kim, T. D. Ladd and Y. Yamamoto, *Phys. Rev. X* **2**, 031007 (2012).
35. A. G. Fowler, M. Mariantoni, J. M. Martinis and A. N. Cleland, *Phys. Rev. A* **86**, 032324 (2012).
36. J. Ghosh, A. G. Fowler and M. R. Geller, *Phys. Rev. A* **86**, 062318 (2012).
37. Y. Li, S. D. Barrett, T. M. Stace and S. C. Benjamin, *Phys. Rev. Lett.* **105**, 250502 (2010).
38. K. Fujii and Y. Tokunaga, *Phys. Rev. Lett.* **105**, 250503 (2010).
39. L. Jiang, J. M. Taylor, A. S. Sørensen and M. D. Lukin, *Phys. Rev. A* **76**, 062323 (2007).
40. K. Fujii, T. Yamamoto, M. Koashi and N. Imoto, *arXiv:1202.6588* (2012).
41. Y. Li and S. C. Benjamin, *New Journal of Physics* **14**, 093008 (2012).
42. N. H. Nickerson, Y. Li and S. C. Benjamin, *Nature Comm.* **4**, 1756 (2012).
43. C. Monroe, R. Raussendorf, A. Ruthven, K. R. Brown, P. Maunz, L.-M. Duan and J. Kim, *arXiv:1208.0391* (2012).
44. R. Raussendorf, S. Bravyi and J. Harrington, *Phys. Rev. A* **71**, 062313 (2005).
45. S. D. Barrett, S. D. Bartlett, A. C. Doherty, D. Jennings and T. Rudolph, *Phys. Rev. A* **80**, 062328 (2009).
46. Y. Li, D. E. Browne, L. C. Kwek, R. Raussendorf and T.-C. Wei, *Phys. Rev. Lett.* **107**, 060501 (2011).
47. K. Fujii and T. Morimae, *Phys. Rev. A* **85**, 010304 (2012).
48. K. Fujii, Y. Nakata, M. Ohzeki and M. Murao, *Phys. Rev. Lett.* **110**, 120502 (2013).

QUANTUM ANNEALING WITH ANTIFERROMAGNETIC FLUCTUATIONS FOR MEAN-FIELD MODELS

YUYA SEKI* and HIDETOSHI NISHIMORI

Department of Physics, Tokyo Institute of Technology,
Oh-okayama, Meguro-ku, Tokyo 152-8551, Japan
* E-mail: y-seki@stat.phys.titech.ac.jp

We discuss the efficiency of quantum annealing with antiferromagnetic fluctuations for mean-field models (the infinite-range ferromagnetic model and the Hopfield model) by analyzing those phase diagrams. The phase diagrams are obtained by using the Suzuki-Trotter formula, the static anzats, and the saddle-point method. The results for the ferromagnetic model show that the antiferromagnetic fluctuations let the system evolve only through second-order transitions when the order of interactions is a finite value greater than 3. The results for the Hopfield model with finite patterns are the same as those for the ferromagnetic model. In contrast, the antiferromagnetic fluctuations cannot avoid first-order transitions of the Hopfield model with many patterns, which is a spin-glass model. We thus conclude that the antiferromagnetic fluctuations are not effective for NP-hard problems.

1. Introduction

A central issue of computer science is to devise an algorithm that efficiently solves combinatorial optimization problems.[1] Roughly speaking, the task of combinatorial optimizations is to find the best choice from many options. From a physical point of view, the problems correspond to the problems of finding the ground states of Ising spin systems. Unfortunately, most of the problems are difficult to solve. The difficulty can be understood from the following naive discussion. First, the number of spin configurations increases exponentially with the system sizes. Thus, the exhaust algorithm that checks the all configurations costs exponentially long computational time. In addition, energy landscapes of Ising spin systems generally have complicated shapes; in other words, the energies as functions of spin configurations have many local minima. This means that the steepest descent method fails to find the ground states unless the method starts near the ground states. Consequently, algorithms for combinatorial optimizations need a mechanism to escape the local minima.

Quantum annealing $(QA)^{2-4}$ can avoid the local minima by using quantum fluctuations. The quantum fluctuations are induced by quantum operators that does not commute with the operators corresponding to the energy functions of Ising spin systems. Most studies adopt the transverse-field operator to induce quantum fluctuations. Quantum annealing searches the ground states with decreasing the strength of quantum fluctuations to zero. The dynamics of the systems obeys the Schrödinger equation. Strong quantum fluctuations make the wave functions of the systems spread over the configurational space independently of the shapes of energy functions. The wave functions begin to concentrate around the minima with decreasing the strength. The systems are expected to escape the local minima owing to the tunneling effects. Eventually, the quantum fluctuations vanish, and the wave functions converge to the ground states. Since QA uses quantum effects, QA is a quantum algorithm.

Quantum annealing often outperforms a classical algorithm, simulated annealing (SA).[5] Simulated annealing is a effective and practical algorithm coming from statistical physics. While QA uses quantum fluctuations, SA uses thermal fluctuations. By simulating the thermal equilibrium state of a system with decreasing its temperature, SA searches the ground state of the system. Quantum annealing can reproduces SA by choosing quantum fluctuations properly.[6] Therefore the efficiency of QA is the same as or more than SA. In fact, numerical and analytical studies showed that QA achieves the faster convergence to the ground states than SA for certain models.[3,7]

Farhi *et al.* proposed a special version of QA, quantum adiabatic computation (QAC).[8] The idea of QAC is essentially equivalent to QA at the point that both algorithms use quantum fluctuations. An important difference between the algorithms is that QAC puts an emphasis on the adiabatic evolution of quantum states.

The adiabatic theorem gives the success probability of QAC. The theorem shows that the runtime of QAC have to be much longer than $\Delta_{\min}^{-2}$ (see Sec. 2.3 in Ref.[4]), where $\Delta_{\min}$ is the minimum energy gap between the instantaneous ground state and the first excited state during the time evolution, to obtain the final ground state with a probability close to unity. Hence, QAC can solve a problem efficiently if the minimum energy gap decreases at most polynomially with increasing the number of spins N. The problem is that the calculation of the gaps for large systems is practically impossible; The required memory resource for storing a Hamiltonian grows exponentially as N increases. Furthermore, the analysis for small size systems often misestimates the efficiency of QAC because of finite size effects.

Phase diagram analysis solves the above problem. It is well known that the energy gap vanishes at quantum phase transition points in the thermodynamic limit $N \to \infty$.[9] Hence, the size dependence of the gap in the vicinity of the transition points determines the efficiency of QAC. If a system exhibits a first-order phase transition, the gap usually decays exponentially at the transition point.[10–12] In contrast, second-order phase transitions are associated with polynomially vanishing gaps;[13] This behavior is consistent with the results from the finite-size scaling theory. Thus, the degree of phase transitions determines the necessary runtime of QAC around the phase transition points.

Unfortunately, conventional QAC with transverse field for several systems undergoes first-order transitions. For instance, Jörg et $al.$ showed that the runtime of conventional QAC grows exponentially even for the ferromagnetic p-spin model with $p \geq 3$.[14] Here, p is the degree of interactions of spins. The ferromagnetic p-spin model with transverse field exhibits second-order quantum phase transitions when $p = 2$, and first-order quantum transitions when $p > 2$.

The above results does not necessarily suggest a complete failure of QAC. Another type of quantum fluctuations possibly improves the efficiency. Note that the degree of freedom of quantum fluctuations is a great advantage of QA. In fact, we find that antiferromagnetic fluctuations can avoid the difficulty of the ferromagnetic p-spin model.[15] An ingenious control of quantum fluctuations induced by the transverse field and antiferromagnetic fluctuations let the system evolve only through second-order transitions. No studies investigated the effects of antiferromagnetic fluctuations on the efficiency of QAC. Accordingly, we must study the class of problems that the new method can solve. To this end, we analyze the Hopfield model,[16] whose coupling constants are randomly distributed. The energy landscape of the Hopfield model has a more complicated shape than that of the ferromagnetic p-spin model. Hence, the ground-state search of the Hopfield model is harder than the ferromagnetic p-spin model. In particular, we focus on the Hopfield model with many patterns. This model has a spin-glass phase. Spin-glass phases are considered as a feature of NP-hard problems.[1]

We investigate the efficiency of the new approach for the ferromagnetic p-spin model from two aspects: the phase diagram and the energy gap.

The phase diagrams are analyzed by using the Suzuki-Trotter formula,[17] the static anzats, and the saddle-point method (see, e.g., Appendix A.1 of Ref.[18]). These are standard strategies to obtain phase diagrams of quantum mean-field systems. We use numerical diagonalization to calculate the energy gap.

We analyze the phase diagram of the Hopfield model. The analysis is almost the same as that of the ferromagnetic p-spin model. The configurational average is taken by using the replica method.[16] We assume the replica symmetry, namely the overlap between replicas is independent of replica indices.

As a result, we find that the phase transitions from a quantum disordered phase to ordered phases does not hamper the QA process. However, phase transitions between a spin-glass phase and a certain ordered phase are always first order. This result indicates that QA with antiferromagnetic fluctuations can solve easy problems, but cannot solve NP-hard problems.

This paper organized as follows. The next section introduce the conventional QA and the new method using antiferromagnetic fluctuations. Section 3 shows the results for the ferromagnetic p-spin model, and Sec. 4 for the Hopfield model. Finally, Sec. 5 is devoted to the summary and conclusion.

2. Quantum annealing

We first introduce conventional QA. Since QA is essentially equivalent to QAC, we use the term QA instead of QAC hereafter. Let us consider the following total Hamiltonian:

$$\hat{H}(t) = s(t)\hat{H}_0 + [1 - s(t)]\hat{V}. \tag{1}$$

Here, $\hat{H}_0$ is a target Hamiltonian whose ground state is desired. Target Hamiltonians are commonly represented in terms of z components of the Pauli matrices $\hat{\sigma}_i^z$ ($i = 1, \ldots, N$). The operator $\hat{V}$ is a driver Hamiltonian that satisfies the following two conditions: (i) it does not commute with the target Hamiltonian, $[\hat{H}_0, \hat{V}] \neq 0$, and (ii) it has an unique trivial ground state. The noncommutativity introduces quantum fluctuations into the system, causing state transitions. A typical driver Hamiltonian is the transverse-field operator $\hat{V}_{\text{TF}} \equiv -\sum_{i=1}^{N} \hat{\sigma}_i^x$ ($i = 1, \ldots, N$). Here, the $\hat{\sigma}_i^x$ are the x component of the Pauli matrices. The time-dependent control parameter $s(t)$ starts at zero and increases monotonically to unity. In other words, the total Hamiltonian varies smoothly from the driver Hamiltonian $\hat{V}$ to the target Hamiltonian $\hat{H}_0$. We assume $s(\tau) = 1$, that is, the running time of QAC is τ. For simplicity, most studies adopt the linear function $s(t) = t/\tau$.

Quantum annealing procedure is as follows: We first prepare the trivial ground state of $\hat{V}$. We then simulate the adiabatic time evolution of the system governed by the Hamiltonian (1). Under the adiabatic condition $\tau \gg \Delta_{\min}^{-2}$, the quantum state stays very close to the instantaneous ground state during the time evolution. We can eventually achieve the ground state of $\hat{H}_0$ at $t = \tau$.

Next, we introduce QA with antiferromagnetic fluctuations. This method uses two driver Hamiltonians:

$$\hat{H}(s, \lambda) = s\{\lambda\hat{H}_0 + (1 - \lambda)\hat{V}_{\mathrm{AFF}}\} + (1 - s)\hat{V}_{\mathrm{TF}}, \tag{2}$$

where

$$\hat{V}_{\mathrm{AFF}} \equiv +N\left(\frac{1}{N}\sum_{i=1}^{N}\hat{\sigma}_i^x\right)^2. \tag{3}$$

The variables s and λ are time-dependent parameters. The initial Hamiltonian has $s = 0$ and any λ; The total Hamiltonian is equal to $\hat{V}_{\mathrm{TF}}$. The final Hamiltonian has $s = \lambda = 1$, and is equals to $\hat{H}_0$. Intermediate values of (s, λ) should be chosen according to the prescription given in the subsequent sections.

It is convenient to consider the quantum annealing procedure on the s-λ plane. A line $\{(s(t), \lambda(t)) \mid 0 \le t \le \tau\}$ is called an annealing path. For example, the line $\lambda = 1$ corresponds to the conventional QA, since the antiferromagnetic term $\hat{V}_{\mathrm{AFF}}$ completely vanishes. We must not take the line $\lambda = 0$ as an annealing path. The total Hamiltonian is diagonalized in the x basis on this line:

$$\hat{H}(s) = s\hat{V}_{\mathrm{AFF}} + (1 - s)\hat{V}_{\mathrm{TF}} = sN\left(\frac{1}{N}\sum_{i=1}^{N}\hat{\sigma}_i^x\right)^2 - (1 - s)\sum_{i=1}^{N}\hat{\sigma}_i^x. \tag{4}$$

Thus, quantum fluctuations completely disappear, and quantum state transitions do not occur. This means that the system does not perform the quantum annealing process. The problem we are concerned with is whether or not we can avoid first-order phase transitions on the line $\lambda = 1$ by taking another annealing path.

3. The ferromagnetic p-spin model

This section analyze the phase diagrams on the s-λ plane and the energy gaps of the ferromagnetic p-spin model. We show that there exist annealing paths to avoid first-order transitions, and that the energy gaps at second-order transitions decrease polynomially as N increases.

The target Hamiltonian of the ferromagnetic p-spin model is given as

$$\hat{H}_0 = -N\left(\frac{1}{N}\sum_{i=1}^{N}\hat{\sigma}_i^z\right)^p,$$ (5)

where p is a integer denoting the order of the interactions. This model has the obvious ground state that the all spins point the $+z$ direction. However, QA with transverse field cannot find the ground state efficiently owing to the first-order quantum phase transition. This model is the simplest hard problem for the QA.

The partition function of the total system is calculated by using the Suzuki-Trotter formula (see Ref.[15] for the detail calculation). This system has two order parameters: magnetization in z direction m^z and x direction m^x. These order parameters have to satisfy the following self-consistent equations:

$$m^z = \frac{ps\lambda(m^z)^{p-1}}{\sqrt{\{ps\lambda(m^z)^{p-1}\}^2 + \{1 - s - 2s(1-\lambda)m^x\}^2}},$$ (6)

$$m^x = \frac{1 - s - 2s(1-\lambda)m^x}{\sqrt{\{ps\lambda(m^z)^{p-1}\}^2 + \{1 - s - 2s(1-\lambda)m^x\}^2}}.$$ (7)

We use the Newton method to find the solutions of Eqs. (6) and (7). The phase diagram is determined by the solution that minimize the pseudo free energy

$$f(s, \lambda; m^z, m^x) = (p-1)s\lambda(m^z)^p - s(1-\lambda)(m^x)^2$$
$$- \sqrt{\{ps\lambda(m^z)^{p-1}\}^2 + \{1 - s - 2s(1-\lambda)m^x\}^2}.$$ (8)

Figure 1 shows the phase diagram of the model for $p = 11$. The order of phase transitions is second-order for small λ. Thus, there exist annealing paths that avoid the first-order transition. We confirmed that second-order boundary remained at least for $5 \le p \le 21$ when $\lambda = 0.1$.

Figure 2 illustrates the minimum energy gap on the line $\lambda = 0.1$. The result shows that the runtime of QA with antiferromagnetic fluctuations increases polynomially.

4. The Hopfield model

This section considers the phase diagram of the Hopfield model.[19,20] The Hopfield model is originally proposed as a prototype of associative memory.

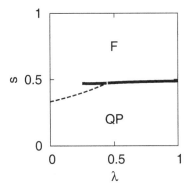

Fig. 1. Phase diagram of the ferromagnetic model with 11-body interactions on the s-λ plane. The solid line represents the first-order transition, and the dashed line the second-order transition. The above region is the ferromagnetic phase, and the below is the quantum paramagnetic phase. The system can evolve only through the second-order transition.

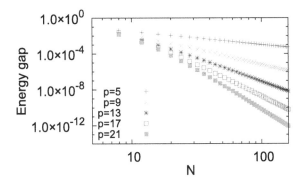

Fig. 2. Minimum energy gap behavior of the ferromagnetic p-spin model on the line $\lambda = 0.1$ for some values of p. The x axis represents the number of spins. Each axis is the logarithmic scale. The energy gaps decrease polynomially up to $N = 160$.

A memory corresponds to a pattern of binary variables $(\xi_1, \ldots, \xi_N)$, where $\xi_i = \pm 1$. We assume that there are p memory patterns, $(\xi_1^\mu, \ldots, \xi_N^\mu)$ for $\mu = 1, \ldots, p$. Note that the definition of p differs from that of the ferromagnetic p-spin model. The Hamiltonian of the Hopfield model is constructed so that the memory patterns are stable states of the system.

First, we discuss the case $p = \mathrm{O}(N^0)$. In order to develop arguments independent of a specific memory pattern, we assume that each binary variable ξ_i^μ takes ± 1 at random. The Hamiltonian embedding p patterns is given as

$$\hat{H}_0 = - \sum_{i_1,\dots,i_k} J_{i_1,\dots,i_k} \hat{\sigma}_{i_1}^z \cdots \hat{\sigma}_{i_k}^z \qquad (9)$$

with

$$J_{i_1,\dots,i_k} = \frac{1}{N^{k-1}} \sum_{\mu=1}^{p} \xi_{i_1}^\mu \cdots \xi_{i_k}^\mu. \qquad (10)$$

Here, k is an integer denoting the order of interactions. The order parameters of this system are the overlaps between the embedded patterns and the state of the system:

$$m_\mu = \frac{1}{N} \sum_{i=1}^{N} \sigma_i^z \xi_i^\mu, \qquad (11)$$

for $\mu = 1, \dots, p$. The phase characterized by the order parameters,

$$m_\mu = \begin{cases} m > 0 & \text{(for a certain pattern)} \\ 0 & \text{(otherwise)} \end{cases}, \qquad (12)$$

is called the retrieval (R) phase. We assume that the first pattern is to be retrieved without loss of generality. The analysis of the phase diagram is almost the same as that of the ferromagnetic p-spin model. The different point is that the resulting self-consistent equations has many solutions such as $(m_1, m_2, \dots) = (m, m, m, 0, \dots)$ and $(m, l, l, l, \dots)$, where m and l are certain positive real numbers in $[0, 1]$. These solutions are called the spurious solutions. Of course, the self-consistent equations have the retrieval solution (12) and the quantum paramagnetic solution. We get the phase diagram by comparing the all free energies corresponding the solutions.

The phase diagram is identical with that of the ferromagnetic p-spin model. The self-consistent equations and the pseudo free energy for the R phase are given by

$$m = \frac{k s \lambda m^{k-1}}{\sqrt{\{k s \lambda m^{k-1}\}^2 + \{1 - s - 2s(1-\lambda)m^x\}^2}}, \qquad (13)$$

$$m^x = \frac{1 - s - 2s(1-\lambda)m^x}{\sqrt{\{k s \lambda m^{k-1}\}^2 + \{1 - s - 2s(1-\lambda)m^x\}^2}}, \qquad (14)$$

and

$$f(s, \lambda; m, m^x) = (k-1)s\lambda m^k - s(1-\lambda)(m^x)^2$$

$$- \sqrt{\{k s \lambda m^{k-1}\}^2 + \{1 - s - 2s(1-\lambda)m^x\}^2}. \qquad (15)$$

We confirmed that spurious solutions have higher free energy than the R solution. Thus, Eqs. (13),(14), and (15) determines the phase diagram. Since the equations are the same as those of the ferromagnetic p-spin model, the resulting phase diagram must be the same. Therefore antiferromagnetic fluctuations are also effective for the problem of finding the ground state of the Hopfield model with finite patterns.

Next, we discuss the case $k = 2$ and $p = 0.04N$. This system has the spin-glass (SG) phase because of the extensive number of the non-retrieval patterns. The phase diagram is shown in Fig. 3. We cannot avoid the first-order transition between the SG phase and the R phase.

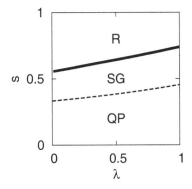

Fig. 3.   Phase diagram of the Hopfield model with many patterns ($p = 0.04N$). The system includes two-body interactions. The solid line represents the first-order transition, and the dashed line the second-order transition. The boundary between the QP phase and the SG phase is always second order. In contrast, the boundary between the SG phase and the R phase is of first order.

5. Summary and conclusion

We investigated the effects of antiferromagnetic fluctuations on the efficiency of QA. We introduced antiferromagnetic fluctuations, and proposed the new approach to QA using antiferromagnetic fluctuations in Sec. 2.

Section 3 studied the efficiency of the approach for the ferromagnetic p-spin model, which cannot be solved by the conventional QA efficiently. We have found that QA with antiferromagnetic fluctuations can avoid first-order phase transitions at least for $5 \leq p \leq 21$. It is expected that QA with antiferromagnetic fluctuations is an efficient algorithm as long as p is a finite integer greater than 3. Although the ferromagnetic p-spin model is a simple problem, antiferromagnetic fluctuations achieve an exponential speedup.

Section 4 analyze the phase diagram of the Hopfield model including many-body interactions. The phase diagram of the Hopfield model with finite patterns indicated that QA with antiferromagnetic fluctuations can find the ground state efficiently. On the other hand, the ground-state search of the Hopfield model with many patterns is hard for QA with antiferromagnetic fluctuations; we cannot avoid the first-order transition between the SG phase and the R phase.

We conclude that the new approach is only effective for phase transitions from the QP phase to the ordered phases (the F phase or the R phase). We cannot avoid the first-order transition between the ordered phases (the SG phase and the R phase) using antiferromagnetic fluctuations. Improvement of QA for models that exhibit phase transitions between ordered phases is a problem to be investigated in the future.

Acknowledgments

Y.S. is grateful for financial supports provided through Research Fellowships of the Japan Society for the Promotion of Scientists.

References

1. A. K. Hartmann and M. Weigt, *Phase Transitions in Combinatorial Optimization Problems: Basics, Algorithms and Statistical Mechanics* (Wiley-VCH, Weinheim, 2005).
2. T. Kadowaki and H. Nishimori, Phys. Rev. E **58**, 5355 (1998).
3. A. Das and B. K. Chakrabarti, Rev. Mod. Phys. **80**, 1061 (2008).
4. V. Bapst, L. Foini, F. Krzakala, G. Semerjian, and F. Zamponi, Phys. Rep. **523**, 127 (2013).
5. S. Kirkpatrick, C. D. Gelatt Jr., and M. P. Vecchi, Science **220**, 671 (1983).
6. R. D. Somma, C. D. Batista, and G. Ortiz, Phys. Rev. Lett. **99**, 030603 (2007).
7. S. Morita and H. Nishimori, J. Math. Phys. **49**, 125210 (2008).
8. E. Farhi, J. Goldstone, S. Gutmann, J. Lapan, A. Lundgren, and D. Preda, Science **220**, 671 (1983).
9. S. Sachdev, *Quantum Phase Transitions* (Cambridge University Press, Cambridge, 1999).
10. T. Jörg, F. Krzakala, J. Kurchan, and A. C. Maggs, Phys. Rev. Lett. **101**, 147204 (2008).
11. A. P. Young, S. Knysh, and V. N. Smelyanskiy, Phys. Rev. Lett. **104**, 020502 (2010).
12. T. Jörg, F. Krzakala, G. Semerjian, and F. Zamponi, Phys. Rev. Lett. **104**, 207206 (2010).

13. R. Schützhold and G. Schaller, Phys. Rev. A **74**, 060304 (2006).
14. T. Jörg, F. Krzakala, J. Kurchan, A. C. Maggs, and J. Pujos, Euro. Phys. Lett. **89**, 40004 (2010).
15. Y. Seki and H. Nishimori, Phys. Rev. E. **85**, 051112 (2012).
16. H. Nishimori, *Statistical Physics of Spin Glasses and Information Processing* (Oxford University Press, Oxford, 2001).
17. M. Suzuki, Prog. Theor. Phys. **56**, 1454 (1976).
18. H. Nishimori and G. Ortiz, *Elements of Phase Transitions and Critical Phenomena* (Oxford University Press, Oxford, 2011).
19. H. Nishimori and Y. Nonomura, J. Phys. Soc. Jpn. **65**, 3780 (1996).
20. E. Gardner, J. Phys. A: Math. Gen. **20**, 3453 (1987).

A METHOD TO CHANGE PHASE TRANSITION NATURE — TOWARD ANNEALING METHODS

RYO TAMURA

International Center for Young Scientists, National Institute for Materials Science,
1-2-1, Sengen, Tsukuba-shi, Ibaraki, 305-0047, Japan
E-mail: tamura.ryo@nims.go.jp
http://www.nims.go.jp/icys/ryo_tamura/tamura_home_e.html

SHU TANAKA

Department of Chemistry, University of Tokyo,
7-3-1, Hongo, Bunkyo-ku, Tokyo, 113-0033, Japan
E-mail: shu-t@chem.s.u-tokyo.ac.jp
http://www.shutanaka.com/

In this paper, we review a way to change nature of phase transition with annealing methods in mind. Annealing methods are regarded as a general technique to solve optimization problems efficiently. In annealing methods, we introduce a controllable parameter which represents a kind of fluctuation and decrease the parameter gradually. Annealing methods face with a difficulty when a phase transition point exists during the protocol. Then, it is important to develop a method to avoid the phase transition by introducing a new type of fluctuation. By taking the Potts model for instance, we review a way to change the phase transition nature. Although the method described in this paper does not succeed to avoid the phase transition, we believe that the concept of the method will be useful for optimization problems.

Keywords: Phase transition; Annealing method; Potts model; Invisible state

1. Introduction

Development of methods to solve optimization problems has been definitely a central issue in science. Optimization problems are spread in a wide area of science such as mathematics, physics, chemistry, biology, and information science.[1–4] Moreover, since optimization problems relate to phenomena in real world and daily life, development of useful optimization methods contributes to growth of industry. Typical examples of optimization problems are designing of transportation system and that of integrated circuit.

Optimization problems are expressed by mathematically well-defined models. In terms of mathematics, the goal of optimization problems is to find $\mathbf{x}^* := \arg\min_{\mathbf{x}} f(\mathbf{x})$, where $f(\mathbf{x})$ is a real-valued function and called cost function. Here, $\mathbf{x}^*$ is referred to as the best solution. When the cost function $f(\mathbf{x})$ is defined by a simple form, we can easily differentiate $f(\mathbf{x})$ and immediately obtain $\mathbf{x}^*$. In general, however, since $f(\mathbf{x})$ is a complicated function in optimization problems, to obtain $\mathbf{x}^*$ directly is difficult. Then, we should develop a method to obtain the best solution of optimization problems. There are many types of optimization problems. Depending on individual types of optimization problems, many efficient but specialized algorithms have been developed mainly in information science.[5]

As mentioned above, to solve optimization problems corresponds to find the state $\mathbf{x}^*$ which minimizes the cost function $f(\mathbf{x})$. As will be shown later, we can relate a cost function to a Hamiltonian of spin system in most cases. In terms of physics, to solve optimization problems is to find the ground state of the corresponding Hamiltonian. Then, to obtain the best solution of optimization problems, we can use methods developed in physics. A generic algorithm was proposed in the context of physics, which imitates natural phenomena. The most famous one is called simulated annealing.[6-14] "Annealing" is a technical terminology in materials science. Annealing is a gradual cooling process of metal alloys and glassy materials to remove stress and defects after these materials are synthesized. The simulated annealing imitates the annealing in computer simulation, which is the origin of the terminology. In the simulated annealing, temperature is introduced into optimization problems as thermal fluctuation. In principle, the best solution can be obtained by decreasing the temperature gradually.[15] Since the simulated annealing is easy to implement, it has been often used in many optimization problems.

There is another typical fluctuation in physics – quantum fluctuation. Annealing method in which quantum fluctuation is controlled was also proposed. This method is called quantum annealing.[16-28] In the quantum annealing, a quantum field which represents quantum fluctuation effect is introduced into optimization problems, and we gradually decrease the quantum field. In principle, the best solution of optimization problems can be obtained as well as the simulated annealing.[29,30] In fact, a quantum field in the quantum annealing plays a similar role with the temperature in the simulated annealing. Since the quantum annealing is easy to implement as the simulated annealing, it has been expected to be an alternative method to the simulated annealing. Efficiency of the quantum annealing has been demonstrated in respective optimization problems.

Annealing methods such as the simulated annealing and quantum annealing seem to be efficient in general. However, there is a serious crisis in annealing methods. It becomes difficult to obtain the best solution by these methods if a phase transition point exists in the process of annealing. Then, it is indispensable to develop a way to avoid the phase transition in optimization problems. In other words, we should discover an annealing process in which no phase transition point exists.

In order to control phase transition behavior on demand, we first should establish a microscopic mechanism to change nature of phase transition. To achieve the issue, we focus on the Potts model which has been used for analysis of phase transition with discrete symmetry breaking.[31,32] The Potts model is a fundamental model in statistical physics and a straightforward generalization of the Ising model.

In this paper, we review a method to change nature of phase transition toward annealing methods. The rest of this paper is organized as follows. In Sec. 2, we review optimization problems with discrete variables and show relation between optimization problems and discrete spin systems which are typical models in statistical physics. In Sec. 3, we explain annealing methods which have developed in physics and have been used to solve optimization problems. In Sec. 4, nature of phase transition is considered in a general way. In Sec. 5, we review properties and phase transition behavior of the Potts model with invisible states. In the Potts model with invisible states, the order of phase transition is changed by controlling the number of invisible states. Section 6 is devoted to conclusion and future perspective.

2. Optimization problems

Optimization problems relate to many real-world problems which are concerned with maximizing benefit or minimizing cost. As stated in Sec. 1, to solve optimization problems is to find the best solution $\mathbf{x}^* := \arg \min_{\mathbf{x}} f(\mathbf{x})$. The cost function of most optimization problems with discrete variables can be represented by Hamiltonian of discrete spin systems such as the Ising model and its generalizations.

Here we explain how to express the traveling salesman problem using the Ising model. The traveling salesman problem is a typical optimization problem with discrete variables.[33–39] In the traveling salesman problem, the complete lists of cities and distances between two cities are given. Let N and $\ell_{i,j}$ be the number of cities and the distance between the i-th and

j-th cities ($1 \leq i, j \leq N$), respectively. By definition, $\ell_{i,j} = \ell_{j,i}$. The traveling salesman problem is to find the shortest path under the following two conditions. The first one is that a traveller can pass through an individual city just one time. The second one is that a traveller finally returns to the initial city. In other words, the start point is the same as the end point. The cost function of traveling salesman problem is the length of path, which is represented by

$$\mathcal{H} = \sum_{a=1}^{N} \ell_{c_a, c_{a+1}},\tag{1}$$

where c_a denotes the city where a traveller passes through at the a-th step. Because of the second condition and $\ell_{i,j} = \ell_{j,i}$, we can choose the initial city arbitrary and $c_{N+1} = c_1$ should be satisfied. Then, the traveling salesman problem is to find $\{c_a\}_{a=1}^{N}$ such that the cost function $\mathcal{H}$ has the minimum value. To express the cost function using a Hamiltonian of a discrete spin model, we introduce a new variable $n_{i,a}(= 0, 1)$ which represents the state of the i-th city at the a-th step. When a traveller passes through the i-th city at the a-th step, $n_{i,a} = 1$ whereas $n_{i,a} = 0$ when a traveller passes through other city at the a-th step. The first condition of traveling salesman problem can be represented by

$$\sum_{a=1}^{N} n_{i,a} = 1, \qquad \forall i(= 1, \cdots, N).\tag{2}$$

Obviously, since a traveller passes through only one city in a single step, the following condition should be satisfied:

$$\sum_{i=1}^{N} n_{i,a} = 1, \qquad \forall a(= 1, \cdots, N).\tag{3}$$

Then, the cost function given by Eq. (1) is rewritten by

$$\mathcal{H} = \sum_{a=1}^{N} \sum_{i,j} \ell_{i,j} n_{i,a} n_{j,a+1}.\tag{4}$$

This cost function can be expressed by the Ising variable as

$$\mathcal{H} = \frac{1}{4} \sum_{a=1}^{N} \sum_{i,j} \ell_{i,j} \sigma_{i,a}^z \sigma_{j,a+1}^z + \text{const.}, \qquad \sigma_{i,a}^z = \pm 1.\tag{5}$$

Here we used the correspondence between the variable $n_{i,a}$ and the Ising variable $\sigma_{i,a}^z$:

$$n_{i,a} = \frac{1}{2} \left(\sigma_{i,a}^z + 1 \right).\tag{6}$$

Then, the conditions given by Eqs. (2) and (3) are rewritten by

$$\sum_{a=1}^{N} \sigma_{i,a}^{z} = -N + 2, \qquad \forall i (= 1, \cdots, N), \qquad (7)$$

$$\sum_{i=1}^{N} \sigma_{i,a}^{z} = -N + 2, \qquad \forall a (= 1, \cdots, N). \qquad (8)$$

We can represent the cost function of traveling salesman problem by the Hamiltonian of Ising model with inhomogeneous interactions on $N \times N$ spins. The number of microscopic states of this system is $\mathcal{O}(2^{N^2})$. When the number of cities N is small, we can easily obtain the ground state by a brute force. However, since the number of microscopic states exponentially increases with N^2, it is difficult to obtain the ground state of the system for large N.

Here, we focus on traveling salesman problem. As mentioned above, the cost function of most optimization problems can be represented by Hamiltonian of discrete spin systems with inhomogeneous interactions as well as the traveling salesman problem. Then, we often face with the same difficulty to solve optimization problems in general. We should develop an intelligent method to obtain the ground state. In information science, efficient but specialized algorithms have been developed to solve respective optimization problems. On the contrary, a generic algorithm was proposed in terms of physics. The most famous algorithm is simulated annealing which will be explained in the next section.

3. Annealing methods

In order to solve optimization problems, a generic algorithm called simulated annealing was proposed in a physical context.[6,7] In the simulated annealing, we introduce the temperature into optimization problems. Since cost function of most optimization problems can be expressed by Hamiltonian of discrete spin systems, the temperature in optimization problem is well-defined. At high temperatures, all states are realized with almost the same probability. In contrast, at zero temperature, the system should be the ground state. Next we gradually decrease the temperature. In principle, the best solution of optimization problems can be definitely obtained when we decrease the temperature slow enough, which was mathematically proved.[15,40] Any system is guaranteed to converge to the stable state in the limit of infinite time if the temperature is decreased in proportion to inverse of the logarithm of time or slower. Since the simulated annealing is easy to implement, it has been adopted for many optimization problems.

After the proposal of the simulated annealing, an alternative method to the simulated annealing – quantum annealing, was proposed.[17] As mentioned above, the simulated annealing can obtain the best solution of optimization problems by imitating thermal fluctuation effect. In contrast, the quantum annealing uses quantum fluctuation effect which is another fluctuation in nature. In the quantum annealing, we introduce a quantum field into optimization problems. For example, if a cost function of an optimization problem is described by the Ising model, we often introduce the transverse magnetic field as a quantum field. Next we decrease the quantum field gradually. The protocol of the quantum annealing is the same as that of the simulated annealing. Then, the quantum annealing is also easy to implement as well as the simulated annealing. In addition, the best solution of optimization problems can be definitely obtained when we decrease the quantum field slow enough. In Refs. 29 and 30, sufficient conditions for convergence of the quantum annealing were given. The strong ergodicity property is proved in three implementation methods of the quantum annealing for the transverse Ising model under a power decay of the transverse field. In Ref. 29, the authors considered the cases of the path-integral Monte Carlo method and the Green's function Monte Carlo method. In Ref. 30, the case of real-time Schrödinger equation was considered. The latter study is based on the idea reported in Ref. 41 in which classical-quantum correspondence was proposed. Recently, experimental demonstrations of the quantum annealing have been done.[42–45] In this way, the quantum annealing is expected to be an efficient algorithm to solve optimization problems as well as the simulated annealing.[46–52] In this section, we consider a mechanism of annealing methods from a viewpoint of statistical physics.

3.1. Mechanism of simulated annealing

In the simulated annealing, we gradually decrease the temperature T and obtain the state when the temperature reaches to $T = 0$. To explain a mechanism of the simulated annealing, suppose we consider the Ising model on a square lattice with homogeneous ferromagnetic interaction. The Hamiltonian of the system is given by

$$\mathcal{H} = -J \sum_{\langle i,j \rangle} \sigma_i^z \sigma_j^z, \qquad \sigma_i^z = \pm 1, \tag{9}$$

$T/J = 0$ $T/J = 2.5$ $T/J = 5$

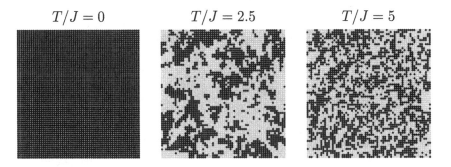

Fig. 1. Black and gray circles indicate +1 and −1 spins, respectively. (Left panel) Perfectly ferromagnetic ordered state. This is the ground state of the model given by Eq. (9). (Middle panel) A typical snapshot of spin configuration in equilibrium state at intermediate temperatures. Short-range ferromagnetic correlation exists. (Right panel) A typical snapshot of spin configuration in equilibrium state at high temperatures. Random spin configuration appears.

where $\langle i, j \rangle$ denotes the nearest-neighbor spin pairs on a square lattice. The ground state of the system is completely ferromagnetic ordered state. It is only necessary to consider the ferromagnetic Ising spin system to show thermal fluctuation effect though the ground state is trivial. Hereafter, the Boltzmann constant is set to unity.

First we consider the equilibrium state of the system on a 64×64 square lattice with periodic boundary condition. At high temperatures, all spins are randomly oriented as shown in the right panel of Fig. 1. As the temperature decreases, short-range ferromagnetic correlation grows, which is shown in the middle panel of Fig. 1. At zero temperature, all spins are parallel, $i.e.$, completely ferromagnetic ordered state shown in the left panel of Fig. 1 appears. As described before, the ground state can be definitely obtained when we decrease the temperature slow enough.[15] This is because the system stays close to the equilibrium at each temperature.

In practice, however, we decrease the temperature with finite speed in the simulated annealing. Then, it is important to consider dynamic nature of the Ising model. There are many types of implementation methods of time evolution. We now focus on the Monte Carlo method which is a stochastic method described in Appendix A. Here we adopt the heat-bath method as the transition probability (see Appendix A). We prepare a random spin configuration depicted in Fig. 2(a) as the initial state. The initial temperature is set to $T_0/J = 10$, which is much higher than the energy scale of magnetic interaction J. Next we decrease the temperature with the

142

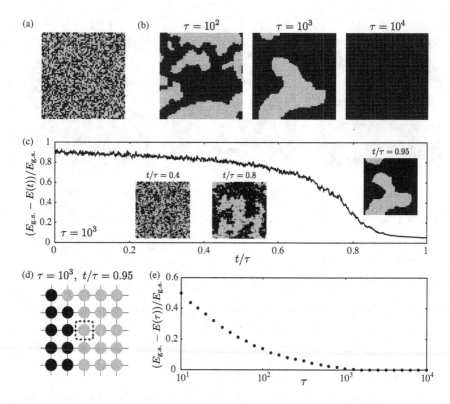

Fig. 2. (a) Initial state. (b) The states at $t = \tau$ where $\tau = 10^2, 10^3, 10^4$. (c) Time development of the internal energy $E(t)$ for $\tau = 10^3$. Snapshots at several t's are also shown. (d) The enlarged view of domain wall. (e) τ-dependence of the internal energy $E(\tau)$.

following schedule:

$$T(t; \tau) = T_0 \left(1 - \frac{t}{\tau}\right), \qquad (0 \le t \le \tau), \qquad (10)$$

where τ^{-1} is the sweeping speed. At $t = \tau$, the temperature becomes zero. Figure 2(b) shows snapshots of spin configuration at $t = \tau$ for various τ's. Here we study the dynamics for $\tau = 10^3$ in detail. To consider the dynamics from a microscopic viewpoint, we calculate the internal energy $E(t)$ at t. To quantify the similarity between present state and the ground state, we consider the quantity: $(E_{\text{g.s.}} - E(t))/E_{\text{g.s.}}$ where $E_{\text{g.s.}}$ is the internal energy of the ground state. Figure 2(c) shows time development of the internal energy and snapshots at several t's for $\tau = 10^3$. As shown in Fig. 2(c), the microscopic state does not almost change at all after $t/\tau = 0.95$. In other words, the state is trapped by the domain wall effect. Here we estimate the probability to break domain walls. The spin indicated by the dotted square in Fig. 2(d) flips with the probability:

$$p_{\text{flip}} = \frac{e^{-2\beta J}}{2\cosh(2\beta J)}, \tag{11}$$

which is very small probability at low temperatures. For example, at $T/J = 0.5$ corresponding to $t/\tau = 0.95$ where the snapshot began to almost stop, $p_{\text{flip}} = 0.00034$. Then it is difficult to break domain wall once the domain wall forms. In order to avoid the domain wall problem, we have to decrease the temperature as slow as possible. Finally, we show τ-dependence of the internal energy $E(\tau)$ obtained by the simulated annealing in Fig. 2(e).

3.2. Mechanism of quantum annealing

In the quantum annealing, we introduce a quantum field and gradually decrease the quantum field at zero temperature. We obtain the state at zero quantum field as the final state. In order to show a mechanism of the quantum annealing, we consider the Ising model with homogeneous ferromagnetic interaction as in the case of the simulated annealing. When the cost function of optimization problem can be described by the Ising model, we often use the transverse field as the quantum field in the quantum annealing. Then the total Hamiltonian is given by

$$\hat{\mathcal{H}} = -J\sum_{\langle i,j\rangle} \hat{\sigma}_i^z \hat{\sigma}_j^z - \Gamma \sum_{i=1}^{N} \hat{\sigma}_i^x, \tag{12}$$

where $\hat{\sigma}_i^\alpha$ denotes the α-component of the Pauli matrix at the site i ($\alpha = x, y, z$). There are many types of implementation methods of the quantum annealing, for example, quantum Monte Carlo simulation, real-time dynamics, and time-dependent density matrix renormalization group (t-DMRG).[53,54] Here we focus on the quantum annealing using the real-time evolution which will be explained in Appendix B.

We consider Γ-dependence of eigenstates and eigenenergies. The ground state depends on the magnitude of transverse field Γ. When $\Gamma = 0$, the ground state is completely ferromagnetic ordered state expressed as $|\uparrow\uparrow \cdots \uparrow\rangle$ or $|\downarrow\downarrow \cdots \downarrow\rangle$. Here $\hat{\sigma}_i^z |\uparrow\rangle = |\uparrow\rangle$ and $\hat{\sigma}_i^z |\downarrow\rangle = -|\downarrow\rangle$. In contrast, the ground state in the limit of $\Gamma \to \infty$ is represented by $|\rightarrow\rightarrow \cdots \rightarrow\rangle$, where $\hat{\sigma}_i^x |\rightarrow\rangle = |\rightarrow\rangle := \frac{1}{\sqrt{2}} (|\uparrow\rangle + |\downarrow\rangle)$. The purpose of the quantum annealing is to obtain the ground state at $\Gamma = 0$. In this case, the ground state at $\Gamma = 0$ is trivial. In general, however, it is difficult to obtain the ground state at $\Gamma = 0$ of the Hamiltonian with inhomogeneous interactions for large N. In contrast, the ground state of the Ising models at $\Gamma \to \infty$ is definitely a trivial state expressed as $|\to\to\cdots\to\rangle$. Then, we can easily prepare the initial state and obtain the ground state at $\Gamma = 0$ by just decreasing transverse field in the quantum annealing.

We calculate eigenenergies of the Hamiltonian given by Eq. (12). Figure 3(a) depicts eigenenergies of the model on 3×3 square lattice with periodic boundary condition. The bold curve in Fig. 3(a) displays Γ-dependence of eigenenergy of the ground state. The curve is smoothly connected between the eigenenergy at large Γ's and that at $\Gamma = 0$. Thus, if we can prepare the ground state at finite Γ as the initial state, we can definitely obtain the ground state at $\Gamma = 0$ in the adiabatic limit.

In practice, however, we decrease the quantum field with finite speed. Then, a nonadiabatic transition occurs during the protocol of the quantum annealing. To show nonadiabatic transition effect, we demonstrate the quantum annealing. The initial transverse field is set to be $\Gamma_0/J = 10$, which is much larger than the scale of magnetic interaction J. We prepare the ground state at Γ_0/J as the initial state. Next we decrease the transverse field with the following schedule:

$$\Gamma(t;\tau) = \Gamma_0 \left(1 - \frac{t}{\tau}\right), \qquad (0 \leq t \leq \tau), \qquad (13)$$

which is the same schedule as Eq. (10). As mentioned above, the ground state at $\Gamma = 0$ is completely ferromagnetic ordered state. The fidelity between the ground state and the state at t obtained by the quantum annealing is calculated. The fidelity is defined by

$$\mathcal{F}(t) := |\langle\psi(t)|\phi_{\text{g.s.}}\rangle|^2, \qquad (14)$$

where $|\psi(t)\rangle$ is the wavefunction at t obtained by the quantum annealing and $|\phi_{\text{g.s.}}\rangle$ is the wavefunction of the ground state at $\Gamma = 0$. When the fidelity closes to 1, the present state obtained by the quantum annealing is similar with the ground state. Figure 3(b) shows the similarity between the present state and the ground state $1 - \mathcal{F}(t)$ as a function of t for some sweeping speeds τ^{-1}. As sweeping speed τ^{-1} increases, $1 - \mathcal{F}(t)$ does not reach to zero. Figure 3(c) shows the sweeping speed τ^{-1}-dependence of the fidelity at $t = \tau$. As the sweeping speed τ^{-1} increases, $1 - \mathcal{F}(\tau)$ increases, which comes from the nonadiabatic transition. Then, in order to avoid the nonadiabatic transition, we have to decrease the quantum field as slow as possible.

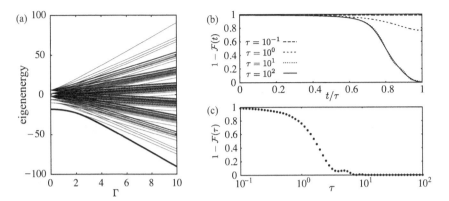

Fig. 3. (a) Γ-dependence of eigenenergies of the ferromagnetic Ising model on square lattice with 3×3 sites. The bold curve indicates the eigenenergy of the ground state. (b) $1 - \mathcal{F}(t)$ as a function of step t for some sweeping speeds τ^{-1}. (c) Sweeping speed τ^{-1}-dependence of $1 - \mathcal{F}(\tau)$.

4. Phase transitions

In Sec. 3, we reviewed mechanisms of the simulated annealing and that of the quantum annealing. In both cases, we introduce a controllable parameter which represents some kind of fluctuation and gradually decrease the parameter. We can prevent unpreferable transition to excited states by decreasing the fluctuation parameter as slow as possible. Then, annealing methods such as the simulated annealing and quantum annealing seem to be versatile for optimization problems. However, we face with difficulties which come from phase transition in annealing methods. As shown above, cost function of most optimization problems with discrete variables can be represented by Hamiltonian of discrete spin systems. According to statistical physics, discrete spin systems exhibit a phase transition in many cases. If there is a transition point in the protocol of annealing methods, it becomes difficult to obtain the best solution.

Phase transitions are divided into two types according to singularity in physical quantities. If the first-order derivative of the free energy is discontinuous, the transition is of the first order and called discontinuous phase transition or first-order phase transition. When the second-order or higher-order derivative of the free energy is discontinuous or divergent, the transition is called continuous transition. When the second-order derivative of the free energy is first discontinuous or divergent, the phase transition is called second-order phase transition. In this section, we explain inherent problem in annealing methods when the system exhibits a phase transition.

4.1. *First-order phase transition*

When a first-order phase transition takes place in the protocol of annealing method, it is difficult to obtain the stable state because of existence of metastable states. As stated above, the first derivative of free energy is discontinuous or divergent at the first-order phase transition point. Typical example of first-order phase transition in nature is ice-water phase transition at 0°C under atmospheric pressure. When we decrease the temperature rapidly under atmospheric pressure, water does not change into ice even below 0°C though the equilibrium state of H_2O is ice below 0°C. This behavior is called supercooled phenomenon. The supercooled phenomenon also appears in many magnetic and electronic compounds when we decrease the temperature rapidly. In these materials, hysteresis curve of the physical quantities such as magnetization obtained by the first derivative of free energy is observed. The hysteresis curve indicates the existence of metastable states. Once the state is trapped in the metastable state, to reach the stable state is difficult.

The same situation happens in theoretical models in which a first-order phase transition occurs. Typical examples of these models are the Blume-Emery-Griffiths model[55] and the Wajnflasz-Pick model.[56] The Hamiltonians of the Blume-Emery-Griffiths model and the Wajnflasz-Pick model are respectively represented by

$$\mathcal{H}_{\mathrm{BEG}} = -J \sum_{\langle i,j \rangle} S_i S_j - J' \sum_{\langle i,j \rangle} S_i^2 S_j^2 - D \sum_i (S_i)^2, \quad S_i = \pm 1, 0, \quad (15)$$

$$\mathcal{H}_{\mathrm{WP}} = -J \sum_{\langle i,j \rangle} S_i S_j, \quad S_i = \underbrace{+1, \cdots, +1}_{g_+}, \underbrace{-1, \cdots, -1}_{g_-}. \quad (16)$$

The former Hamiltonian was proposed in order to explain the phase transition nature of ^3He-^4He mixture.[55] The latter one can analyze the phase transition behavior of spin-crossover materials, and the number of $+1$ states and that of -1 states in the latter Hamiltonian are g_+ and g_-, respectively.[57-63] Furthermore, we can transfer the Wajnflasz-Pick model into the following Hamiltonian at finite temperature T:

$$\mathcal{H}_{\mathrm{WP}} = -J \sum_{\langle i,j \rangle} \sigma_i^z \sigma_j^z - (h - \frac{T}{2} \log \frac{g_+}{g_-}) \sum_i \sigma_i^z, \quad \sigma_i^z = +1, -1. \quad (17)$$

The bias of g_+ and g_- induces the temperature-dependent chemical potential.

These models given by Eq. (15) and Eqs. (16) and (17) are generalized Ising models and exhibit a thermal-induced first-order phase transition for a certain parameter region. When we decrease the temperature rapidly, hysteresis curve appears in these models. Since the cost function of optimization problems with discrete variables can be represented by Hamiltonian of discrete spin systems, a first-order phase transition sometimes occurs in optimization problems. If the supercooled phenomenon occurs in optimization problems because of first-order phase transition, we cannot obtain the best solution definitely. Then, in order to improve annealing method, we should avoid the first-order phase transition point in the protocol of annealing method. In terms of the quantum annealing, difficulty in systems where a first-order phase transition appears can be explained as follows. The energy gap is likely to be exponentially small at the first-order phase transition point, which leads to exponential complexity.[64–66] In Ref. 67, the authors concluded that the exact cover problem, which is a typical optimization problem, in the quantum annealing exhibits a first-order phase transition. Recently, the authors in Ref. 68 studied antiferromagnetic fluctuation effect in the ferromagnetic p-spin model with transverse field with the quantum annealing in mind. Originally, the model exhibits a first-order phase transition. However, they found that we can make a path that avoids the first-order phase transition point.

4.2. Second-order phase transition

When a second-order phase transition occurs in optimization problems, we face with other type of difficulty caused by critical slowing down. As described before, the first derivative of free energy is analytic but the second derivative of free energy is discontinuous or divergent at the second-order phase transition point. Physical quantities can be described by power-law behavior near the second-order phase transition point. Suppose we consider the magnetic system in which a second-order phase transition occurs at $T = T_c$, where T_c is called critical temperature. The specific heat C, magnetization m, and magnetic susceptibility χ near the critical point behave as

$$C(T) \propto |T - T_c|^{-\alpha}, \quad m(T) \propto |T - T_c|^{\beta}, \quad \chi(T) \propto |T - T_c|^{-\gamma}, \quad (18)$$

where α, β, and γ are critical exponents. Each critical exponent does not have an independent value and $\alpha + 2\beta + \gamma = 2$ called the Rushbrooke relation is satisfied. Although the relations given by Eq. (18) are behavior in equilibrium state, a similar relation exists in nonequilibrium process. An order parameter which describes the second-order phase transition reaches to the equilibrium value with exponential decay except at the critical temperature. However, the order parameter reaches to the equilibrium value with a power-law decay at the critical temperature. The relaxation time τ relates to the correlation length ξ. When a second-order phase transition takes place, the relaxation time diverges in the thermodynamic limit, which is represented as

$$\tau \sim \xi^z, \tag{19}$$

where z is called the dynamical critical exponent. A typical example of system in which a second-order phase transition occurs is the ferromagnetic Ising model given by Eq. (9). The ferromagnetic Ising model is the simplest model and exhibits order-disorder transition with spontaneous symmetry breaking.

The critical slowing down relates to the domain-wall problem called the Kibble-Zurek phenomena.[69,70] The performance of annealing methods has been studied in terms of the Kibble-Zurek mechanism.[71-73] Since the cost function of optimization problems with discrete variables can be represented by Hamiltonian of discrete spin systems, a second-order phase transition sometimes occurs as well as a first-order phase transition. Then, in order to make annealing methods more efficient, we should avoid the second-order phase transition point in the protocol of annealing method.

5. Potts model with invisible states

In order to change phase transition nature with fixing a symmetry which breaks at the transition point, a new discrete spin model called Potts model with invisible states was proposed.[74] In this section, we explain phase transition behavior of the model. In condensed matter physics and materials science, when we find a phase transition with discrete symmetry breaking in real materials or complicated theoretical models, we often analyze the phase transition nature using the Potts model.[32] The Potts model is a cornerstone of discrete spin models in statistical physics. In fact, the analysis using the Potts model succeeded in many cases.[75-94]

Suppose we consider a phase transition with q-fold symmetry breaking in d dimension. In order to investigate the phase transition, we often refer the phase transition nature of ferromagnetic q-state Potts model on d-dimensional lattice. The Hamiltonian of the model is given by

$$\mathcal{H} = -J \sum_{\langle i,j \rangle} \delta_{\sigma_i, \sigma_j}, \qquad \sigma_i = 1, \cdots, q, \quad (q \in \mathbb{N}), \tag{20}$$

where the sum is over pairs of nearest-neighbor spins on d-dimensional lattice. The Potts model is a generalized Ising model since the model is equivalent to the Ising model when $q = 2$. The ferromagnetic Potts model exhibits a phase transition at finite temperature in d dimension $(d \geq 2)$. Phase transition nature depends on the number of states q and the spacial dimension d. For example, in two dimension, when $q \leq 4$, a second-order phase transition occurs whereas a first-order phase transition occurs when $q > 4$. In both cases, q-fold symmetry breaks at the phase transition point. Not only the order of phase transition but also the critical phenomena were investigated. In many cases, nature of phase transition observed in experiment and obtained in complicated theoretical model can be explained by the ferromagnetic Potts model. Recently, however, phase transitions where the behavior is different from the ferromagnetic Potts model. For instance, a first-order phase transition with threefold symmetry breaking was found in two-dimensional frustrated systems,[95-97] although the three-state ferromagnetic Potts model in two dimension exhibits a second-order phase transition with threefold symmetry breaking. As shown above, in the ferromagnetic Potts model, when the number of states q and the spatial dimension d are given, the order of the phase transition is determined. Then, unconventional phase transitions such as the abovementioned examples cannot be represented by the ferromagnetic Potts model.

In order to overcome the fact, a generalized Potts model called Potts model with invisible states was proposed.[74] The Hamiltonian is given by

$$\mathcal{H} = -J \sum_{\langle i,j \rangle} \delta_{s_i, s_j} \sum_{\alpha=1}^{q} \delta_{s_i, \alpha}, \qquad s_i = 1, \cdots, q+r, \quad (q+r \in \mathbb{N}). \quad (21)$$

If and only if $1 \leq s_i = s_j \leq q$, interaction between the i-th and j-th sites works. Then, the state $s_i (\geq q+1)$ is regarded as redundant states. Here the redundant states are called invisible states. In the ground state, all spins have the same value from 1 to q and q-fold symmetry is broken. When $r = 0$, the model represented by Eq. (21) is equivalent to the standard Potts model given by Eq. (20). Thus, the model is a straightforward generalization of the Potts model.

In order to clarify the effect of invisible states, let us show another representation of the Potts model with invisible states when $r \geq 1$:

$$\mathcal{H} = -J \sum_{\langle i,j \rangle} \delta_{\sigma_i, \sigma_j} \sum_{\alpha=1}^{q} \delta_{\sigma_i, \alpha} - T \ln r \sum_{i} \delta_{\sigma_i, 0}, \qquad \sigma_i = 0, 1, \cdots, q. \quad (22)$$

It should be noted that the Hamiltonian given by Eq. (22) is the same as that given by Eq. (21) at each temperature. This fact can be confirmed by comparing partition functions of both models. Note that the spin s_i in the Hamiltonian given by Eq. (21) takes from 1 to $q + r$ whereas the spin σ_i in the Hamiltonian given by Eq. (22) takes from 0 to q. The invisible states are labeled by $\sigma_i = 0$ in Eq. (22). The second term in the Hamiltonian given by Eq. (22) represents temperature-dependent chemical potential of invisible states, which is similar with the Wajnflasz-Pick model given by Eq. (17).

Here we consider two-spin systems of the Potts model with invisible states given by Eq. (21) comparing with the standard ferromagnetic Potts model given by Eq. (20). In the standard ferromagnetic Potts model, the ground-state energy is $-J$ and the energy of excited states is 0, which is the same as the Potts model with invisible states. In both models, the number of ground states is q. However the number of excited states in each model is different. The number of excited states in the standard ferromagnetic Potts model is $q^2 - q$ whereas that in the Potts model with invisible states is $q^2 - q + 2qr + r^2$. Thus, the number of excited states increases due to existence of invisible states. The increase of the number of excited states affects nature of phase transition.

Next we explain nature of phase transition in the Potts model with invisible states. In Refs. 74, 98, and 99, the authors investigated phase transition behavior of the Potts model with invisible states in two dimension for $q \leq 4$ and large r. If there is no invisible states ($r = 0$), a second-order phase transition occurs in the model for $q \leq 4$. The authors calculated temperature dependences of specific heat and order parameter which detects the q-fold symmetry breaking by Monte Carlo simulations. As the temperature decreases, the order parameter becomes non-zero value at the temperature where the specific heat has the maximum value. These behaviors suggest an existence of phase transition. In order to determine the order of the phase transition, probability distribution of internal energy at the temperature where the specific heat has the maximum value was calculated. The bimodal distribution was observed, which is a characteristic behavior of the first-order phase transition. Moreover, the finite-size scaling analysis of the first-order phase transition was performed. In the finite-size scaling analysis, the authors found that the latent heat remains in the thermodynamic limit. From the above results, a first-order phase transition occurs in the Potts model with invisible states in two dimension for large r even when $q \leq 4$. In addition, the authors confirmed that as r increases, the transition temperature decreases but the latent heat increases.

The results obtained by Monte Carlo simulations suggest that the invisible states play a role to change phase transition nature. In Refs. 74, 98, and 99, to confirm the fact, the authors also studied the phase transition of the Potts model with invisible states by the Bragg-Williams approximation which is a kind of mean-field analysis. The Bragg-Williams approximation of the standard ferromagnetic Potts model given by Eq. (20) concludes that a second-order phase transition occurs when $q = 2$ whereas a first-order phase transition occurs when $q \geq 3$.[100] Here we explain the Bragg-Williams approximation of the Potts model with invisible states. For convenience, we use the representation of Hamiltonian given by Eq. (22). Let x_α be the fraction of the α-th state $(0 \leq \alpha \leq q)$. Obviously, $\sum_{\alpha=0}^{q} x_\alpha = 1$ is satisfied. Here $\alpha = 0$ indicates the invisible state. Since now we consider the case that the q-fold symmetry breaks at the transition point, one of q-states is selected in the ferromagnetically ordered phase. The label of the selected state is set to $\alpha = 1$. Then the fractions are given by

$$x_0 = t, \tag{23}$$

$$x_1 = \frac{1}{q}(1-t)\left[1 + (q-1)s\right], \tag{24}$$

$$x_\alpha = \frac{1}{q}(1-t)(1-s), \qquad (2 \leq \alpha \leq q), \tag{25}$$

where $0 \leq s, t \leq 1$. Then the internal energy E^{BW} and the entropy S^{BW} in the Bragg-Williams approximation are expressed as

$$
\begin{aligned}
E^{\mathrm{BW}}(s,t) &= -\frac{zJ}{2}\sum_{\alpha=1}^{q} x_\alpha^2 - x_0 T \ln r \\
&= -\frac{zJ(1-t)^2}{2q}\left[(q-1)s^2 + 1\right] - tT\ln r,
\end{aligned} \tag{26}
$$

$$
\begin{aligned}
S^{\mathrm{BW}}(s,t) &= -\sum_{\alpha=0}^{q} x_\alpha \ln x_\alpha \\
&= -t\ln t - (1-t)\left[\frac{1+(q-1)s}{q}\ln\frac{1+(q-1)s}{1-s}\right. \\
&\quad \left. + \ln\frac{(1-t)(1-s)}{q}\right].
\end{aligned} \tag{27}
$$

Then, the free energy is given by

$$F^{\mathrm{BW}}(s,t) = E^{\mathrm{BW}}(s,t) - TS^{\mathrm{BW}}(s,t). \tag{28}$$

By analyzing the free energy, we can obtain the transition temperature and latent heat. As mentioned above, the Bragg-Williams approximation concludes that when there are no invisible states, a second-order phase transition occurs for $q = 2$ and a first-order phase transition occurs for $q \geq 3$. Then we first focus on the case of $q = 2$. When $(q, r) = (2, 1), (2, 2)$, and $(2,3)$, a second-order phase transition with twofold symmetry breaking occurs. In contrast, when $q = 2$ and $r \geq 4$, a first-order phase transition occurs. In addition, the authors confirmed that when $q \geq 3$, a first-order phase transition occurs regardless of r. Furthermore, as r increases, the transition temperature decreases but the latent heat increases. Then, the authors in Refs. 74, 98, and 99 concluded that the invisible states play a role to change to a first-order phase transition from a second-order phase transition. On other words, invisible states enlarge the latent heat and prevent the ordering. After the authors proposed the model, the phase transition nature of the model on various lattices was investigated by analytical calculations.[101–105]

Finally, we consider the structure of interaction in the Potts model with invisible states. We show another representation of the Potts model with invisible states. In the representation, the interaction tensor is used.[99] We first consider the standard ferromagnetic Potts model. Let $\vec{S}_i$ be a q-dimensional binary vector. $\vec{S}_i$ represents the microscopic state in the i-th site. Only one of elements in the vector is unity whereas the other elements are zero. The position of unity indicates the state, $e.g.$, $\vec{S}_i = {}^{\mathrm{T}}(0, 1, 0, \cdots, 0)$ means that the state of the i-th spin is the second state. Here the symbol T is the transpose of vector. By using the vector representation, the Hamiltonian of standard ferromagnetic Potts model is given by

$$\mathcal{H} = -J \sum_{\langle i,j \rangle} \delta_{s_i, s_j} = -\sum_{\langle i,j \rangle} {}^{\mathrm{T}}\vec{S}_i \hat{J} \vec{S}_j, \qquad s_i = 1, \cdots, q, \qquad (29)$$

where $\hat{J}$ is a $q \times q$ diagonal matrix:

$$\hat{J} = \mathrm{diag}(J, J, \cdots, J). \qquad (30)$$

In a similar way, we can represent the Hamiltonian of Potts model with invisible states. The Hamiltonian using the vector representation is given by

$$\mathcal{H} = -J \sum_{\langle i,j \rangle} \delta_{t_i, t_j} \sum_{\alpha=1}^{q} \delta_{t_i, \alpha} = -\sum_{\langle i,j \rangle} {}^{\mathrm{T}}\vec{T}_i \hat{J} \vec{T}_j, \qquad t_i = 1, \cdots, q + r, \quad (31)$$

where $\vec{T}_i$ is a $(q+r)$-dimensional binary vector and $\hat{J}$ is a $(q+r) \times (q+r)$ diagonal matrix:

$$\hat{J} = \mathrm{diag}(\underbrace{J, \cdots, J}_{q}, \underbrace{0, \cdots, 0}_{r}). \tag{32}$$

From a viewpoint of interaction structure, the results obtained in previous studies can be summarized as follows. The order of phase transition can be changed by just expanding the space of the microscopic state. The invisible states correspond to zero elements in the interaction tensor. Some unconventional phase transitions found in two-dimensional frustrated systems can be represented by the Potts model with invisible states, which is an important progress in statistical physics and condensed matter physics. Unfortunately, the method to change the nature of phase transition is not efficient for annealing methods since the order of phase transition is only changed in this method. However, the method explained in this section is just a simple extension of the Potts model. There are many remaining degrees of freedom, *e.g.*, off-diagonal elements in the interaction tensor. Then, we believe that we can avoid a phase transition by employing a similar strategy.

6. Conclusion and future perspective

In this paper, we reviewed a method to change nature of phase transition toward annealing methods. The annealing methods such as the simulated annealing and quantum annealing are regarded as an efficient general technique to solve optimization problems widely. In Sec. 2, as an example of optimization problems, we introduced the traveling salesman problem. The traveling salesman problem can be represented by the Ising model. In this way, most optimization problems with discrete variables can be represented by Hamiltonian of discrete spin systems. Then, we can use generic algorithms proposed in terms of physics – annealing method, to solve optimization problems.

In the simulated annealing, we introduce the temperature into optimization problems and gradually decrease the temperature. On the other hand, in the quantum annealing, the quantum field such as a transverse field in the Ising model is introduced, and the quantum field is decreased. The best solution of optimization problems can be definitely obtained when we decrease the temperature or quantum field slow enough. In Sec. 3, mechanisms of the simulated annealing and quantum annealing were explained by using the Ising model on a square lattice as an example.

Although the annealing methods are versatile methods for optimization problems, we face with difficulties which come from phase transition in the annealing procedure. We explained the difficulties when the first-order phase transition or second-order phase transition occur in the optimization problems in Sec. 4. Then, in order to improve annealing method more efficient, we should avoid the phase transition in the annealing methods. In Sec. 5, we showed a way to change nature of phase transition in the Potts model by introducing a new type of fluctuation called invisible state. The invisible state is redundant state, and the ground state in the Potts model with invisible states does not change that in the standard ferromagnetic Potts model. The authors in Refs. 74, 98, and 99 concluded that the invisible states play a role to change to a first-order phase transition from a second-order phase transition. Then, the method using the invisible states does not succeed to avoid the phase transition which induces the difficulty to obtain the best solution of optimization problems. However, study on phase transition nature in the Potts model with invisible states is just getting started. It is an important issue to explore inherent properties of invisible states. In addition, extensions of the Potts model with invisible states are interesting, which was explained in the end of Sec. 5. Moreover, in view of optimization problems, investigation of the effect of invisible states in spin systems with inhomogeneous interactions is significant.

A way to change nature of phase transition using invisible states is easy to implement for optimization problems as well as the temperature and quantum field which are used in typical annealing methods. Then, we strongly believe that underlying concept presented in this paper will be useful annealing methods to solve optimization problems.

Acknowledgments

The authors would like to thank Mikio Nakahara for giving us excellent opportunities to write a lecture note on quantum annealing last year and to present our study in Kinki University two years ago. One of the authors R.T. first met Hiroaki Matsueda and Jun-ichi Inoue in the symposium on interface between quantum information and statistical physics. In addition, R.T. got to know Keisuke Fujii bonding over the lecture note. After that, he continues to communicate with them. These are quite precious experiences for us. Thus, it is our great pleasure and honor indeed to celebrate Mikio Nakahara's 60th birthday. The authors agree with Mikio Nakahara who said that I hope all of us meet again in the year 2072, Mikio Nakahara's next KANREKI.

The authors are also grateful to Naoki Kawashima, Jie Lou, Yoshiki Matsuda, Seiji Miyashita, Takashi Mori, Yohsuke Murase, Taro Nakada, Masayuki Ohzeki, Per Arne Rikvold, Takafumi Suzuki, Yusuke Tomita, and Eric Vincent for their valuable comments. R.T. is partially supported by Grand-in-Aid for Scientific Research (C) (25420698) and National Institute for Materials Science (NIMS). S.T. is partially supported by Grand-in-Aid for JSPS Fellows (23-7601). The computations in the present work were performed on computers at the Supercomputer Center, Institute for Solid State Physics, University of Tokyo.

Appendix A. Monte Carlo method

In Sec. 3.1, we demonstrated the simulated annealing using the Monte Carlo method. In this appendix, we show how to implement the Monte Carlo method. Suppose we consider the Ising model with inhomogeneous interactions. The Hamiltonian is given by

$$\mathcal{H} = -\sum_{\langle i,j \rangle} J_{ij}\sigma_i^z\sigma_j^z, \qquad (\sigma_i^z = \pm 1). \tag{A.1}$$

The procedure of Monte Carlo method is as follows:

Step 1 We prepare an initial state.

Step 2 We choose a spin randomly.

Step 3 We calculate the local energy at the chosen site i. The local energy is defined by

$$h_i^{(\mathrm{eff})} := \sum_{j\,(\mathrm{n.n.\ of\ }i)} J_{ij}\sigma_j^z, \tag{A.2}$$

where the summation is over the nearest-neighbor sites of the i-th site. Note that the Hamiltonian can be represented using the local energy:

$$\mathcal{H} = -\frac{1}{2}\sum_i h_i^{(\mathrm{eff})}\sigma_i^z. \tag{A.3}$$

Step 4 We flip the chosen spin according to probability by some way. In general, the probability can be calculated by the local energy, which will be explained later.

Step 5 We continue the procedure from Step 2 to Step 4.

There are two famous decision rules of the probability. One is called the heat-bath method which is given by

$$p_{\mathrm{HB}}(\sigma_i^z \to -\sigma_i^z) = \frac{\mathrm{e}^{-\beta h_i^{(\mathrm{eff})} \sigma_i^z}}{2\cosh(\beta h_i^{(\mathrm{eff})})}. \tag{A.4}$$

The other is called the Metropolis method which is given by

$$p_{\mathrm{M}}(\sigma_i^z \to -\sigma_i^z) = \begin{cases} 1 & (h_i^{(\mathrm{eff})}\sigma_i^z < 0) \\ \mathrm{e}^{-2\beta h_i^{(\mathrm{eff})}\sigma_i^z} & (h_i^{(\mathrm{eff})}\sigma_i^z \geq 0) \end{cases}. \tag{A.5}$$

Both of them satisfy the detailed balance condition. However, the detailed balance condition is just a sufficient condition for stochastic process toward equilibrium state. Then, a decision rule of the probability without detailed balance condition was proposed.[106–108] Using the method, we can obtain the stable state efficiently. Recently, a mechanism of the method has been studied in terms of nonequilibrium statistical physics.[109–112] In the simulated annealing, we decrease the temperature during the procedure from Step 2 to Step 5.

Appendix B. Real-time dynamics by Schrödinger equation

In Sec. 3.2, we demonstrated the quantum annealing based on real-time dynamics. In this appendix, we explain how to calculate real-time dynamics. We first consider time-independent Hamiltonian. The Schrödinger equation is given by

$$i\frac{\partial}{\partial t}\left|\psi(t)\right\rangle = \hat{\mathcal{H}}\left|\psi(t)\right\rangle, \tag{B.1}$$

where the Planck constant $\hbar$ is set to unity. The time evolution of wave function is expressed as

$$\left|\psi(t)\right\rangle = \mathrm{e}^{-i\hat{\mathcal{H}}t}\left|\psi(t=0)\right\rangle =: \hat{U}(t)\left|\psi(t=0)\right\rangle, \tag{B.2}$$

where $\hat{U}(t)$ is the time-evolution operator. For time-independent Hamiltonians, we can immediately obtain the wavefunction at time t if we assign the time t and the initial wave function $\left|\psi(t=0)\right\rangle$. In order to compute the time-evolution operator, the Hamiltonian should be diagonalized. Let $\hat{\mathcal{U}}$ be unitary matrix which diagonalizes the Hamiltonian $\hat{\mathcal{H}}$. Then,

$$\hat{\mathcal{H}}_{\mathrm{d}} = \hat{\mathcal{U}}^\dagger \hat{\mathcal{H}} \hat{\mathcal{U}} = \mathrm{diag}(\epsilon_1, \cdots, \epsilon_{\mathcal{D}}), \tag{B.3}$$

where $\mathcal{D}$ is the number of microscopic states. For $S = 1/2$ spin system with N sites, $\mathcal{D} = 2^N$. By using the unitary matrix $\hat{\mathcal{U}}$, the time-evolution operator is given by

$$U(t) = e^{-i\hat{\mathcal{H}}t} = \hat{\mathcal{U}}e^{-i\hat{\mathcal{H}}_d t}\hat{\mathcal{U}}^\dagger. \tag{B.4}$$

Since the matrix $\hat{\mathcal{H}}_d$ is a diagonal matrix, the matrix exponential is tractable:

$$e^{-i\hat{\mathcal{H}}_d t} = \mathrm{diag}(e^{-i\epsilon_1 t}, \cdots, e^{-i\epsilon_D t}). \tag{B.5}$$

Next we consider the case that the Hamiltonian depends on time. In this case, the Schrödinger equation is given by

$$i\frac{\partial}{\partial t}|\psi(t)\rangle = \hat{\mathcal{H}}(t)|\psi(t)\rangle. \tag{B.6}$$

The time evolution of wave function is formally described as

$$|\psi(t)\rangle = \hat{\mathcal{T}}\exp\left[-i\int_0^t dt'\,\hat{\mathcal{H}}(t')\right]|\psi(t=0)\rangle, \tag{B.7}$$

where $\hat{\mathcal{T}}$ is the time-ordered product of operators. In the quantum annealing, we introduce a quantum field and decrease gradually the quantum field. Then, we can obtain the time evolution of wave function by calculating Eq. (B.7).

References

1. G. Strang, *Introduction to Applied Mathematics* (Wellesley-Cambridge Press, 1986).
2. J. C. Miller and J. N. Miller, *Statistics for Analytical Chemistry* (Ellis Horwood Ltd, 1993).
3. O. C. Martin, R. Monasson, and R. Zecchina, *Theor. Comp. Sci.* **265**, 3 (2001).
4. A. K. Hartmann and M. Weigt, *Phase Transitions in Combinatorial Optimization Problems* (Wiley-VCH, 2005).
5. J. J. Moré and S. J. Wright, *Optimization Software Guide (Frontiers in Applied Mathematics)* (Society for Industrial and Applied Mathematics, 1987).
6. S. Kirkpatrick, C. D. Gelatt Jr., and M. P. Vecchi, *Science* **220**, 671 (1983).
7. S. Kirkpatrick, *J. Stat. Phys.* **34**, 975 (1984).
8. L. Davis (ed.), *Genetic Algorithms and Simulated Annealing* (Pitman Publishing, 1987).
9. P. J. M. van Laarhoven and E. H. L. Aarts, *Simulated Annealing: Theory and Applications, Mathematics and Its Applications Vol. 37* (Springer, 1987).
10. H. Szu and R. Hartley, *Phys. Lett. A* **122**, 157 (1987).
11. E. Aarts and J. Korst, *Simulated Annealing and Boltzmann Machines: A Stochastic Approach to Combinatorial Optimization and Neural Computing (Wiley Series in Discrete Mathematics & Optimization)* (Wiley, 1989).
12. L. Ingber, *Math. Comp. Modeling* **18**, 29 (1993).

158

13. W. L. Goffe, G. D. Ferrier, and J. Rogers, *Journal of Econometrics* **60**, 65 (1994).
14. M. de S. G. Tsuzuki (ed.), *Simulated Annealing - Advances, Applications and Hybridizations* (InTech, 2012).
15. S. Geman and D. Geman, *IEEE Transactions on Pattern Analysis and Machine Intelligence* **6**, 721 (1984).
16. A. B. Finnila, M. A. Gomez, C. Sebenik, C. Stenson, and J. D. Doll, *Chem. Phys. Lett.* **219**, 343 (1994).
17. T. Kadowaki and H. Nishimori, *Phys. Rev. E* **58**, 5355 (1998).
18. J. Brooke, D. Bitko, T. F. Rosenbaum, and G. Aeppli, *Science* **284**, 779 (1999).
19. E. Farhi, J. Goldstone, S. Gutmann, J. Lapan, A. Lundgren, and D. Preda, *Science* **292**, 472 (2001).
20. G. E. Santoro, R. Martoňák, E. Tosatti, and R. Car, *Science* **295**, 2427 (2002).
21. A. Das and B. K. Chakrabarti, *Quantum Annealing and Related Optimization Methods* (Springer, 2005).
22. D. A. Battaglia and L. Stella, *Contemporary Physics* **47**, 195 (2006).
23. G. E. Santoro and E. Tosatti, *J. Phys. A: Math. Gen.* **39**, R393 (2006).
24. A. Das and B. K. Chakrabarti, *Rev. Mod. Phys.* **80**, 1061 (2008).
25. M. Ohzeki and H. Nishimori, *J. Comp. Theor. Nanoscience* **8**, 963 (2011).
26. D. de Falco and D. Tamascelli, *RAIRO - Theoretical Informatics and Applications* **45**, 99 (2011).
27. S. Suzuki, J. Inoue, and B. K. Chakrabarti, *Quantum Ising Phases and Transitions in Transverse Ising Models (Lecture Note in Physics Vol. 862)* (Springer, 2012).
28. V. Bapst, L. Foini, F. Krzakala, G. Semerjian, and F. Zamponi, *Phys. Rep.* **523**, 127 (2013).
29. S. Morita and H. Nishimori, *J. Phys. A* **39**, 13903 (2006).
30. S. Morita and H. Nishimori, *J. Phys. Soc. Jpn.* **76**, 064002 (2007).
31. R. B. Potts, *Proc. Cambridge Philos. Soc.* **48**, 106 (1952).
32. F. Y. Wu, *Rev. Mod. Phys.* **54**, 235 (1982).
33. G. Dantzig, R. Fulkerson, and S. Johnson, *Journal of the Operations Research Society of America* **2**, 393 (1954).
34. S. Lin, *Bell System Technical Journal* **44**, 2245 (1965).
35. E. L. Lawler, J. K. Lenstra, A. H. G. R. Kan, and D. B. Shmoys, *The Traveling Salesman Problem: A Guided Tour of Combinatorial Optimization (Wiley Series in Discrete Mathematics & Optimization)* (Wiley, 1985).
36. D. S. Johnson and L. A. McGeoch, *Local Search in Combinatorial Optimisation*, eds. E. H. L. Aarts and J. K. Lenstra (Wiley, 1997), p. 215.
37. M. Dorigo, *IEEE Transaction on Evolutionary Computation* **1**, 53 (1997).
38. G. Gutin and A. P. Punnen (eds.), *The Traveling Salesman Problem and Its Variations* (Springer, 2007).
39. D. L. Applegate, R. E. Bixby, V. Chvátal, and W. J. Cook, *The Traveling Salesman Problem: A Computational Study (Princeton Series in Applied Mathematics)* (Princeton University Press, 2007).

40. E. Aarts and J. Korst, *Simulated Annealing and Boltzmann Machines: A Stochastic Approach to Combinatorial Optimization and Neural Computing* (Wiley, 1984).

41. R. D. Somma, C. D. Batista, and G. Ortiz, *Phys. Rev. Lett.* **99**, 030603 (2007).

42. M. W. Johnson, M. H. S. Amin, S. Gildert, T. Lanting, F. Hamze, N. Dickson, R. Harris, A. J. Berkley, J. Johansson, P. Bunyk, E. M. Chapple, C. Enderud, J. P. Hilton, K. Karimi, E. Ladizinsky, N. Ladizinsky, T. Oh, I. Perminov, C. Rich, M. C. Thom, E. Tolkacheva, C. J. S. Truncik, S. Uchaikin, J. Wang, B. Wilson, and G. Rose, *Nature* **473**, 194 (2011).

43. A. P. -Ortiz, N. Dickson, M. D. -Brook, G. Rose, and A. A. -Guzik, *Scientific Reports* **2**, 571 (2012).

44. S. Boixo, T. Albash, F. M. Spedalleri, N. Chancellor, and D. A. Lidar, *Nat. Commun.* **4**, 2067 (2013).

45. N. G. Dickson, M. W. Johnson, M. H. Amin, R. Harris, F. Altomare, A. J. Berkley, P. Bunyk, J. Cai, E. M. Chapple, P. Chavez, F. Cioata, T. Cirip, P. deBuen, M. Drew-Brook, C. Enderud, S. Gildert, F. Hamze, J. P. Hilton, E. Hoskinson, K. Karimi, E. Ladizinsky, N. Ladizinsky, T. Lanting, T. Mahon, R. Neufeld, T. Oh, I. Perminov, C. Petroff, A. Przybysz, C. Rich, P. Spear, A. Tcaciuc, M. C. Thom, E. Tolkacheva, S. Uchaikin, J. Wang, A. B. Wilson, Z. Merali, and G. Rose, *Nat. Commun.* **4**, 1903 (2013).

46. K. Kurihara, S. Tanaka, and S. Miyashita, *Proceedings of the 25th Conference on Uncertainty in Artificial Intelligence* (2009).

47. I. Sato, K. Kurihara, S. Tanaka, H. Nakagawa, and S. Miyashita, *Proceedings of the 25th Conference on Uncertainty in Artificial Intelligence* (2009).

48. S. Tanaka, R. Tamura, I. Sato, and K. Kurihara, *Kinki University Quantum Computing Series Vol. 5* (World Scientific, 2012), p. 169.

49. I. Sato, S. Tanaka, K. Kurihara, S. Miyashita, and H. Nakagawa, *Neurocomputing* **121**, 523 (2013).

50. R. Martoňák, G. E. Santoro, and E. Tosatti, *Phys. Rev. E* **70**, 057701 (2004).

51. L. Stella, G. E. Santoro, and E. Tosatti, *Phys. Rev. B* **72**, 014303 (2005).

52. O. Titiloye and A. Crispin, *Discrete Optimization* **8**, 376 (2011).

53. S. Suzuki and M. Okada, *Quantum Annealing and Related Optimization Methods* eds. A. Das and B. K. Chakrabarti (Springer, 2005), p. 207.

54. S. Suzuki and M. Okada, *Interdisciplinary Information Sciences* **13**, 49 (2007).

55. M. Blume, V. J. Emery, and R. B. Griffiths, *Phys. Rev. A* **4**, 1071 (1971).

56. J. Wajnflasz and R. Pick, *J. Phys. Colloq. France* **32**, C1 (1971).

57. R. Zimmermann, *J. Phys. Chem. Sol.* **44**, 151 (1983).

58. A. Bousseksou, J. Nasser, J. Linares, K. Boukheddaden, and F. Varret, *J. Phys. I France* **2**, 1381 (1992).

59. A. Hauser, J. Jeftić, H. Romstedt, R. Hinek, and H. Spiering, *Coord. Chem. Rev.* **190-192**, 471 (1999).

60. K. Boukheddaden, I. Shteto, B. Hôo, and F. Varret, *Phys. Rev. B* **62**, 14806 (2000).

61. S. Miyashita and N. Kojima, *Prog. Theor. Phys.* **109**, 729 (2003).

62. M. Nishino, S. Miyashita, and K. Boukheddaden, *J. Chem. Phys.* **118**, 4594 (2003).

63. H. Tokoro, S. Miyashita, K. Hashimoto, and S. Ohkoshi, *Phys. Rev. B* **73**, 172415 (2006).

64. B. Altshuler, H. Krovi, J. Roland, *arXiv:* 0908.2782.

65. M. H. S. Amin and V. Choi, *Phys. Rev. A* **80**, 062326 (2009).

66. B. Altshulera, H. Krovib, and J. Roland, *Proc. Natl. Acad. Sci. USA* **107**, 12446 (2010).

67. A. P. Young, S. Knysh, and V. N. Smelyanskiy, *Phys. Rev. Lett.* **104**, 020502 (2010).

68. Y. Seki and H. Nishimori, *Phys. Rev. E* **85**, 051112 (2012).

69. T. W. B. Kibble, *J. Phys.* **A9**, 1387 (1976).

70. W. H. Zurek, *Nature* **317**, 505 (1985).

71. T. Caneva, R. Fazio, and G. E. Santoro, *Phys. Rev. B* **76**, 144427 (2007).

72. G. Biroli, L. F. Cugliandolo, and A. Sicilia, *Phys. Rev. E* **81**, 050101(R) (2010).

73. S. Suzuki, *J. Phys.: Conf. Ser.* **302**, 012046 (2011).

74. R. Tamura, S. Tanaka, and N. Kawashima, *Prog. Theor. Phys.* **124**, 381 (2010).

75. M. Weger and I. B. Goldberg, *Solid State Phys.* **28**, 1 (1973).

76. N. Szabo, *J. Phys. C: Solid State Phys.* **8**, L397 (1975).

77. D. Kim and R. J. Joseph, *J. Phys. A: Math. Gen.* **8**, 891 (1975).

78. D. Mukamel, M. E. Fisher, and E. Domany, *Phys. Rev. Lett.* **37**, 565 (1976).

79. A. Aharony, K. A. Müller, and W. Berlinger, *Phys. Rev. Lett.* **38**, 33 (1977).

80. M. Bretz, *Phys. Rev. Lett.* **38**, 501 (1977).

81. E. Domany, M. Schick, and J. S. Walker, *Phys. Rev. Lett.* **38**, 1148 (1977).

82. B. Barbara, M. F. Rossignol, and P. Bak, *J. Phys. C: Solid State Phys.* **11**, L183 (1978).

83. A. N. Berker, S. Ostlund, and F. A. Putnam, *Phys. Rev. B* **17**, 3650 (1978).

84. E. Domany and E. K. Riedel, *J. Appl. Phys.* **42**, 1315 (1978).

85. B. K. Das and R. B. Griffiths, *J. Chem. Phys.* **70**, 5555 (1979).

86. E. Domany and M. Schick, *Phys. Rev. B* **20**, 3828 (1979).

87. D. Blankschtein and A. Aharony, *J. Phys. C: Solid State Phys.* **13**, 4635 (1980).

88. D. Blankschtein and A. Aharony, *Phys. Rev. B* **22**, 5549 (1980).

89. M. J. Tejwani, O. Ferreira, and O. E. Vilches, *Phys. Rev. Lett.* **44**, 152 (1980).

90. J. F. Gouyet, *Ordering in Two Dimensions* ed. S. K. Sinha (North-Holland, 1980), p. 355.

91. J. F. Gouyet, B. Sapoval, and P. Pfeuty, *J. Phys. Lett.* **41**, L115 (1980).

92. R. L. Park, T. L. Einstein, A. R. Kortan, and L. D. Roelofs, *Ordering in Two Dimensions* ed. S. K. Sinha (North-Holland, 1980), p. 17.

93. D. Blankschtein and A. Aharony, *J. Phys. C: Solid State Phys.* **14**, 1919 (1981).

94. E. Domany, Y. Shnidman, and D. Mukamel, *J. Phys. C: Solid State Phys.*

15, L495 (1982).

95. R. Tamura and N. Kawashima, *J. Phys. Soc. Jpn.* **77**, 103002 (2008).

96. S. Okumura, H. Kawamura, T. Okubo, and Y. Motome, *J. Phys. Soc. Jpn.* **79**, 114705 (2010) .

97. R. Tamura and N. Kawashima, *J. Phys. Soc. Jpn.* **80**, 074008 (2011).

98. S. Tanaka, R. Tamura, and N. Kawashima, *J. Phys.: Conf. Ser.* **297**, 012022 (2011).

99. R. Tamura, S. Tanaka, and N. Kawashima, *Kinki University Series on Quantum Computing Vol. 7* (World Scientific, 2012), p. 217.

100. T. Kihara, Y. Midzuno, and T. Shizume, *J. Phys. Soc. Jpn.* **9**, 681 (1954).

101. A. C. D. van Enter, G. Iacobelli, and S. Taati, *Prog. Theor. Phys.* **126**, 983 (2011).

102. A. C. D. van Enter, G. Iacobelli, and S. Taati, *Rev. Math. Phys.* **24**, 1250004 (2012).

103. T. Mori, *J. Stat. Phys.* **147**, 1020 (2012).

104. D. A. Johnston and R. P. K. C. M. Ranasinghe, *J. Phys. A: Math. Theor.* **46**, 225001 (2013).

105. N. Ananikian, N. S. Izmailyan, D. A. Johnston, R. Kenna, and R. P. K. C. M. Ranasinghe, *J. Phys. A: Math. Theor.* **46**, 385002 (2013).

106. H. Suwa and S. Todo, *Phys. Rev. Lett.* **105**, 120603 (2010).

107. H. Suwa and S. Todo, *arXiv:* 1207.0258.

108. H. Suwa and S. Todo, *Monte Carlo Methods and Applications: Proceedings of the 8th IMACS Seminar on Monte Carlo Methods, August 29 - September 2, 2011, Borovets, Bulgaria* eds. K. K. Sabelfeld and I. Dimov (De Gruyter, 2012) p. 213.

109. J. Shi, T. Chen, B. Yuan, and P. Ao, *arXiv:* 1206.2189.

110. M. Ohzeki and A. Ichiki, *arXiv:* 1307.0434.

111. A. Ichiki and M. Ohzeki, *Phys. Rev. E* **88**, 020101(R) (2013).

112. Y. Sakai and K. Hukushima, *J. Phys. Soc. Jpn.* **82**, 064003 (2013).

COMPUTATIONAL ANALYSIS OF THE FIRST STAGE OF THE PHOTOSYNTHETIC SYSTEM, THE LIGHT-DEPENDENT REACTION, BY QUANTUM CHEMICAL SIMULATION METHOD

MASAHITO TADA-UMEZAKI

Institute of Natural Medicine, University of Toyama, Sugitani 2630, Toyama, 930194, Japan
E-mail: masume@inm.u-toyama.ac.jp

Calculation was done about the optimized geometries of special pair consisting of "bacteriochlorophyll a" based on the PDB structural data (3zuw) by quantum simulation method. The quantum chemical calculation for the optimized geometries exhibited the electronic excitation transition with large oscillator strength at 860 nm, characterizing P870. Therefore, the optimized geometries were reasonably assigned to a model of real P870. The geometries of cationic dimer were also optimized by the above method. The charge separation process upon photo-excitation was concluded to be impossible by considering the difference between the heat of formation of the neutral and cation dimers and the ionization energy of the neutral dimer.

1. Introduction

A chloroplast in the body of the green plant is performed, but, as for the photosynthesis, the ability to run photosynthesis normal with the chinning liquid that grind the leaf of the plant, and took a chloroplast is lost.[1] But it became to be succeeded in adjusting the flawless chloroplast authentic sample which can perform CO2 fixation later (Jensen & Bassham, 1966[2]). It shows that it is not effective for the study of the photosynthesis function to apply the commonplaceness-like analysis, such as extracts the ingredient material and analyzes of chemistry mechanically. It may be said that the study of Jensen & Bassham was the important discovery that led a photosynthesis study at a point suggesting that structure characteristics with the constitution factor are essential to the ingredient factors such as the molecules that a photosynthetic function constitutes it only existing, and not appearing, the appearance of the function for a new stage.

In this study, the development of the adjustment method of the chloro-plast authentic sample went up the new stairs, and the photosynthesis study accomplished qualitative evolution to the isolation of the photosynthesis unit. One top was success of the X-ray crystallographic analyses of the photosynthesis unit.[3] It occured successively as a result of Rhodobactor sphaeroides, and an X-ray crystallographic analysis result[4] of Rhodopseudomonas viridis was announced in 1985, and the photosynthesis study acquires new knowledge qualitatively.

In late years it began to be recognized, "quantum mechanics has an important meaning in the photosynthesis study" to a photosynthesis researchers.[5] The quantum mechanics is function in an electronic state that described a system to develop a photosynthesis based on a quantum theory so that quantum mechanics to a quantum theory contribute to a photosynthesis study, and it is necessary to consider physics quantity observed based on the function. The function can be found in condition describing state of the quantum system using the molecular orbital method.[6] If it is two or three atoms system, a highly precise calculation beyond experiment precision is enabled, and the non-empirical molecular orbital method represented by the ab initio method is characterized by that, a small molecule can demand a high state function of the precision. On the other hand, it is difficult that, big molecules apply this numeration without concentrating a laborer such as supposing, and 1SCF calculating a thing and structure to take off a substituent to reduce constitution atomicity in the big system that this study is going to make targeted for a calculation. In contrast, the calculation of most representative elements and transition elements except the rare-earth element is enabled, and semiempirical method (SEMO method) progresses steadily even after entering in the 21st century and evolves for the high molecular orbital method of the versatility. In this study, calculation of structure optimization about a special pair of Rhodobactor sphaeroides using the SEMO method which evolved and consider the process based on a demanded electronic state function in the photosynthesis in photosystem II.

2. Calculation Methods

A special pair (SP) located in the reaction center does optical pumping in the early period of photosynthesis and has electric charge separation through electronic property of nature to follow it or a photochemistry process, and it is thought that a photosynthetic chemistry process starts. This study is intended to describe electronic property of nature process beginning with the optical pumping of the process in the early period of photosynthesis based on the electron state with the quantum theory. It includes one electron process such as optical pumping or the electric charge separation in this process. Therefore, not the RHF approximation that has been usually adopted by the molecular orbital method, it was adopted a UHF approximation. The proper thing is reported recently only after using a UHF approximation for the dissociation process of H2.[7]

The photosynthetic reaction center of Rhodobactor sphaeroides is comprised of photosystem II including the special pair consisting of Bacteriochlorophyll a (BChl a). The photosynthesis reaction of the most important higher plant is forms by the photosynthesis system: photosystem II. Besides, the special pair is comprised of Chlorophyll a. It is essential to calculate that difference of chlorophyll in both systems exactly to elucidate the characteristic of these systems and each role.

As the numeration which satisfied these requests, SCIGRESS MO Compact Pro V1.0.6[8] used in this study. For the structure optimization calculation, making input data of the geometry structure using Z-Matrix as an internal coordinate and calculated the full structure optimization of all coordinates by the PM6 method. As for the structure optimization routine in MOPAC, the EF method became the default, but used the BFGS method which could appoint the convergence precision of the calculation. The energy level that became the standard of the convergence judgment considered the case that converged to a minimum figure to be perfection convergence and did the structure with the optimization geometry structure. It was performed a standard vibration analysis calculation about the optimization structure that completely converged and inspected that the structure existing corresponding to the minimum of the potential curved surface.

MO-S (INDO/S) was used corresponding to a UHF approximation installed in the program for an excited state calculation. The number of item CI which considered is 100. It was adapted Pariser-Parr, Nishimoto - again and chosen it and can use -Weiss, Ono, Ono -Klopman, a Nishimoto - type of DasGupta- Fujinaga to congratulate again for a calculation of 2 center repulsion integral calculus in MO-S. By the excited state calculation of the conventional RHF approximation level, it has been used Nishimoto - most commonly again because an expression of adaption it had good agreement with the experiment. It is appointed a default an expression of used it Nishimoto - again in MO-S. By this calculation of the UHF approximation level, I compared the actual value with the calculated value about chlorophyll, and the correspondence with the actual value used an expression of DasGupta-Fujinaga formula.

3. Results and Discussions

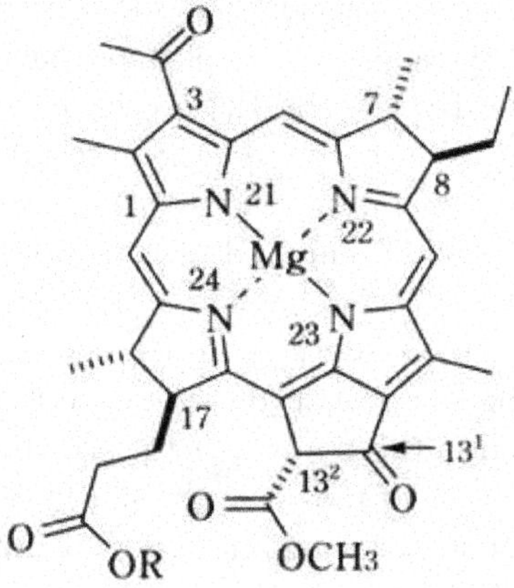

Fig. 1. Structural Formula of Bacteriochlorophyll a, R=Phytyl

The special pair of Rhodobactor sphaeroides is comprised of BChl a. The molecular frame of these molecules is a flat mask, but, judging from the structural formula in Figure 1, the molecules side reflection symmetricalness disappears. According to the structure (PDBID: 3zuw) that was found for the X-ray crystallographic analysis (Figure 2), two BChl take structure of face to face.

From the mutual orientation of the molecular axis, they can be classifed dimeric structure in the type V that inclined to an H type and the V-shape that molecules axis almost has orientation of the parallelism from N22-Mg-N24 axis. The molecular axial crossing-over corner expressed it as a crossing-over corner of the molecules axis that I reflected in two molecular plane interfaces here. The initial structure to converge in SP of the type V is limited to the considerably small range, and an H type is almost provided from the initial structure of the wide range except it. It is showed optimization geometry structure to belong to these two kinds of models in Figure 3.

A series of numberings are accomplished to distinguish an atom in Z-

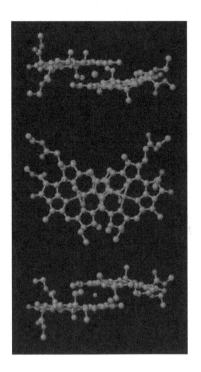

Fig. 2. Observed Geometry of Special Pair (PDBID: 3zuw).

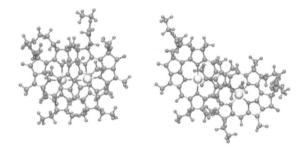

Fig. 3. Optimized Geometry of Special Pair

Matrix used for the description of the internal coordinate, and to appoint it. The indication of the structure in Figure 3 performs one or two and a numbering of Mg, N21 according to a promise of Z-Matrix and puts atom 1-2 on the X-axis and displays 1-2-3 by a method to put on XY plane.

To an electronic appendix, it is showed the input data which used for the optimization geometry structure calculation in Figure 3.

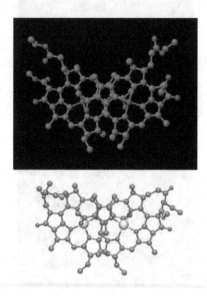

Fig. 4. Comparison of V-SP to PDB Data (3zuw).

Figure 4 showed the comparison of PDB result and the calculation result to how well structure found by a structure optimization calculation could reproduce an actual survey result to judge it. It showed the structure of the same viewpoint that I reflected in the interface of 2 molecules in the figure to facilitate a comparison. The hydrogen atoms are not displayed by the result of the X-ray crystallographic analysis. Therefore it assumes the indication of the calculation result hydrogen atoms non-indication. The correspondence of the structural characteristic of both is very good and understands that the result of the structure optimization calculation can reproduce structure of the SP satisfactorily.

Figure 5 showed the result of the standard vibration calculation. It was inspected in the negative vibration corresponding to the imaginary mode being included in neither structure that these structure was the structure that could exist that could be located all in the minimum of the potential curved surface.

About these two kinds of dimers, it was showed Mg-Mg atomic distance and the molecular axial crossing-over corner in Table 1 for quantity char-

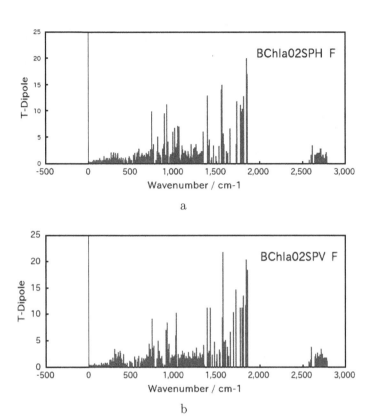

Fig. 5. Results of Normal Mode Calculations of Geometry Optimized Special Pairs, a: H -SP, b: V -SP

acterizing the dimeric structure other than generation energy. The minus sign of the molecular axial crossing-over corner shows that a direction of the heap objects than the direction of the PDB result. There are a few differences between H-SP and V-SP, but V-SP is slightly low and knows that it is with most stability structure when looking at the value of the generation energy. The Mg-Mg atomic distance in V-SP becomes considerably smaller than an actual value.

It showed the result of the INDO/S calculation by the excited state calculation of the UHF approximation level in Figure 6.

An S2 level corresponding to the singlet transition showed only 0 transition. The characteristic absorption transition of the visible level is located to 597nm with 845nm, the H type in V-SP. 845nm of V-SP. It can be

Table 1. Representative Molecular Constants for Special Pairs

Type of SP	Heat of Formation ΔH/kJ mol $-$ 1	Mg-Mg Distance r/nm	Cross Angle* θ/o
PDBID: 3zuw		0.783	37
SP			
H-type	-2085.95	0.488	-10
V-type	-2086.48	0.619	48
SP cation			
HC-type	-1496.10	0.490	
VC-type	-1525.67	0.614	

Note: * Cross Angle between Molecular Axes

conclude that these belong to absorption transition characterizing P870. Generally, it is thought that this agreement should be considerably satisfied when considering that the agreement with the laboratory finding of the excited state calculation result. Because the calculated type V dimer reproduces structure of the SP of Rb. sphaeroides well and has excitation transition of 845nm that is characteristic absorption of P870, it is thought that this calculation is reasonable.

Furthermore, we calculated about cationic state (the thing that one electron jumped out) of the P870 special pair to check "how light energy transfer in a reaction center". We calculated the structure optimization for the dimeric cation based on the neutral dimeric optimization structure of the P870 special pair. It can be considered that electric charge separation following optical pumping or the cationic generation to be a process of known photoionization or the light ionization conventionally, when energy of the light is bigger than the ionization energy of the system. Highest Occupied Molecular Orbital (HOMO) energy of cationic P870 by this calculation was -7.75eV. This supports energy of the light of the vacuum ultraviolet area of 160nm. Therefore, photoionization of the P870 to the light ionization is impossible energetically by the optical pumping of the visible light.

Energy differences between P870 cation and normal state were found with 561 kJ/mol. This energy difference supports energy of the light of the ultraviolet area of approximately 210nm. This result does not contradict the above-mentioned conclusion that the light ionization of the special pair by the visible light is impossible energetically.

Because excitation transition of 845nm correspondence to for characteristic absorption of P870, it is concluded when this dimer demanded by a calculation is the model dimer which can belong to substance of P870.

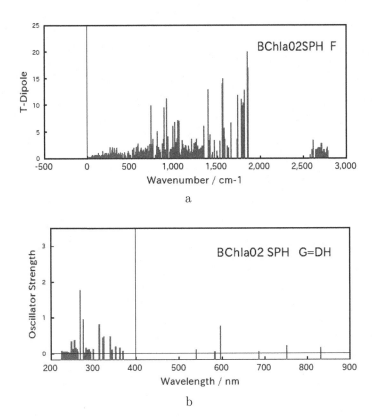

Fig. 6. The Estimated Spectra of Excitation Transition, a:V-SP, b:H-SP.

4. Conclusion

We performed a structure optimization calculation of P870 comprised of BChl a using the UHF approximation method. This dimer has big excitation transition of the oscillator strength to 845nm from UHF approximate INDO/S calculation. Using the same method, we performed the structure optimization calculation of the P870 cation of BChl a, and the convergence structure of the P870 cation was found. HOMO energy of cationic P870 normal state and the energy differences between P870 cation and normal state showed that the function structure of the supermolecule that transmission of the light energy was carried out altogether between the photosynthesis pigment molecules of the reaction center is existed.

References

1. F. Fujishige, "Photosynthesis" (Tokyo University publication society, 1882).
2. R. Jensen and J. Bassham, *Proc. Natl. Acad. Sci. USA.* **56**, 1095 (1966).
3. J. Deisenhofer, O. Epp, K. Miki, R. Huber and H. Michel, *Nature* **318**, 618 (1985).
4. J. P. Allen et al., *Proc. Natl. Acad. Sci. USA* **83**, 8589 (1986).
5. K. Sonoike, "Kougousei to wa nanika" (Kodansha, 2008).
6. E. Lewars, Computational Chemistry (Kluwer Academic Publishers, 2003).
7. A. Sagan and U. Nagashima, *J. Comput. Chem. Jpn.* **10**(1), 44 (2011).
8. SCIGRESS MO Compact Version 1.0.6 Professional, Copyright Fujitsu Limited, 1997-2011.

TWO-QUBIT GATE OPERATION ON SELECTED NEAREST NEIGHBORING QUBITS IN A NEUTRAL ATOM QUANTUM COMPUTER

ELHAM HOSSEINI LAPASAR[1], KENICHI KASAMATSU[2,3], SÍLE NIC CHORMAIC[4], TAKEJI TAKUI[1], YASUSHI KONDO[2,3], MIKIO NAKAHARA[2,3] and TETSUO OHMI[2]

[1] *Department of Chemistry and Materials Science, Graduate School of Science, Osaka City University, Sumiyoshi, Osaka 558-8585, Japan.*

[2] *Research Center for Quantum Computing, Interdisciplinary Graduate School of Science and Engineering, Kinki University, Higashi-Osaka, 577-8502, Japan.*

[3] *Department of Physics, Kinki University, Higashi-Osaka, 577-8502, Japan.*

[4] *Light-Matter Interactions Unit, OIST Graduate University, Onna-son, Okinawa 904-0495, Japan*

Quantum information science has rapidly grown with the promise to build up a quantum computer to solve many classically intractable problems by making use of properties of quantum particles, such as superposition and entanglement. Systems of trapped neutral atoms is one of the promising candidates for implementing a scalable quantum computer, since neutral atoms have an advantage of an intrinsically weaker interaction with the environment. We have discussed a design of a neutral atom quantum computer with a selective two-qubit gate operation.[1] In this contribution, we propose a feasible experiment towards the selective two-qubit gate operation that is less demanding than our original proposal, although the gate operation is limited between neighboring atoms. We evaluate the process of a two-qubit gate operation that is applied on nearest neighboring trapped atoms and estimate the upper bound of the gate operation time and corresponding gate fidelity. This proposal can be demonstrated using current technology, which would be a starting point toward the realization of a fully controlled selective two-qubit gate operation using neutral atoms.[2]

References

1. E. Hosseini Lapasar, K. Kasamatsu, Y. Kondo, M. Nakahara, and T. Ohmi, JPSJ, **80**, 114003 (2011).
2. E. Hosseini Lapasar, K. Kasamatsu, S. Nic Chormaic, T. Takui, Y. Kondo, M. Nakahara, and T. Ohmi, arXiv:1310.6112.

A SIMPLE OPERATOR QUANTUM ERROR CORRECTION SCHEME AVOIDING FULLY CORRELATED ERRORS

CHIARA BAGNASCO

Interdisciplinary Graduate School of Science and Engineering, Kinki University,
3-4-1 Kowakae, Higashi-Osaka, Osaka, 577-8502, Japan

YASUSHI KONDO and MIKIO NAKAHARA

Department of Physics, Kinki University,
3-4-1 Kowakae, Higashi-Osaka, Osaka, 577-8502, Japan

We propose a simple three-qubit OQEC scheme avoiding fully correlated noise. One data qubit is protected by encoding it with two arbitrary ancillae, which can be in the uniformly mixed state. We demonstrated our scheme experimentally with a three-qubit NMR quantum computer. Experimental results show that our scheme eliminates the effects of fully correlated noise.

Keywords: NMR; QEC; Quantum Computing.

1. Introduction

One of the key obstacles to effectively implementing quantum computation is the quantum system's vulnerability to external noise. Quantum error correction (QEC) strategies were devised to protect quantum information against interaction with the environment. Well-known examples include *quantum error correcting codes* (QECC), *noiseless subsystems* (NS), and *decoherence-free subspaces* (DFS).

An umbrella approach known as *Operator Quantum Error Correction* (OQEC), which generalizes and unifies the above techniques, was later introduced by Kribs *et al.*[1] In fact, as we shall see below, any given quantum error correction scheme is found to be a special case of OQEC determined by specifying the triple $\{\mathcal{R}, \mathcal{E}, \mathfrak{A}\}$.

We here discuss a simple OQEC scheme with three qubits, in which one data qubit is protected by encoding it with two ancillae whose state can be quite arbitrary. Our proposal provides the simplest possible NS, in terms of the number of CNOT gates, that is robust against fully correlated noise defined by the errors $\{E_k\} = \{\sqrt{p_1}\,\sigma_0^{\otimes 3}, \sqrt{p_2}\,\sigma_x^{\otimes 3}, \sqrt{p_3}\,\sigma_y^{\otimes 3}, \sqrt{p_4}\,\sigma_z^{\otimes 3}\}$, with $\sum p_i = 1$. Here σ_0 denotes the identity matrix, and σ_x, σ_y, σ_z are the Pauli matrices. Note that, in the following, we use the term NS in the broader sense defined in Ref. 1, which does not necessarily entail the decomposition of the error operators into a direct sum of irreducible representations of $SU(2)$.

2. Operator Quantum Error Correction

Let $\mathcal{H}$ be the Hilbert space for some quantum system. Let

$$\mathcal{H} = \left(\mathcal{H}^A \otimes \mathcal{H}^B\right) \oplus \mathcal{K}$$

with $\dim(\mathcal{H}^A) = m$, $\dim(\mathcal{H}^B) = n$ and $\dim(\mathcal{K}) = \dim(\mathcal{H}) - mn$ be a fixed decomposition of said space. Hereafter, we will discuss the case when $\dim(\mathcal{K}) = 0$. We assume that $\mathcal{H}^A$ and $\mathcal{H}^B$ are spanned by the orthonormal sets $|\alpha_i\rangle$ and $|\beta_j\rangle$, respectively. Let us define the projection operator

$$P_{\mathfrak{A}} = \sum_i |\alpha_i\rangle\langle\alpha_i| \otimes \mathbb{I}^B \oplus 0_{\dim(\mathcal{K})},$$

such that $P_{\mathfrak{A}}(\mathcal{H}) = \mathcal{H}^A \otimes \mathcal{H}^B$, and a super-operator

$$\mathcal{P}_{\mathfrak{A}}(\cdot) = P_{\mathfrak{A}}(\cdot)P_{\mathfrak{A}}.$$

Let $\mathfrak{B}(\mathcal{H})$ be the set of operators on $\mathcal{H}$. Suppose $\mathcal{E}$ is a quantum error channel acting on $\mathfrak{B}(\mathcal{H})$. For any $\sigma \in \mathfrak{B}(\mathcal{H})$, every such channel admits an operator sum representation

$$\mathcal{E}(\sigma) = \sum_k E_k \sigma E_k^\dagger.$$

The operators $\{E_k\} \in \mathfrak{B}(\mathcal{H})$ are known as the Kraus operators (or errors) associated with $\mathcal{E}$. For a given decomposition of $\mathcal{H}$, we define $\mathfrak{A}$ as the operator semigroup in $\mathfrak{B}(\mathcal{H})$ such that

$$\mathfrak{A} = \{\sigma \in \mathfrak{B}(\mathcal{H}) : \sigma = \sigma^A \otimes \sigma^B,$$
$$\text{for some } \sigma^A \in \mathfrak{B}(\mathcal{H}^A) \text{ and } \sigma^B \in \mathfrak{B}(\mathcal{H}^B)\}. \tag{1}$$

The B-sector of $\mathfrak{A}$ is said to be $\mathcal{E}$-correctable with respect to the above decomposition if there exists a trace-preserving quantum recovery operation $\mathcal{R}$ on $\mathfrak{B}(\mathcal{H})$ such that

$$(\mathrm{Tr}_A \circ \mathcal{P}_{\mathfrak{A}} \circ \mathcal{R} \circ \mathcal{E})(\sigma) = \mathrm{Tr}_A(\sigma) \tag{2}$$

for any $\sigma \in \mathfrak{A}$.

Any given quantum error correction scheme is determined by specifying the triple $\{\mathcal{R}, \mathcal{E}, \mathfrak{A}\}$:

When $\mathcal{R} \neq id$ (identity channel) and $\mathfrak{A}$ is a subspace, the scheme is a QECC.

When $\mathcal{R} = id$ and $\mathfrak{A}$'s are a subspace (an algebra), the scheme is a DFS (NS).

3. Simplest NS Under Fully Correlated Noise

Let us consider a *fully correlated* noise where all the qubits are simultaneously affected by the same errors. This may happen, for example, when photons are sent one by one through an optical fiber with a fixed imperfection (assuming the scattering of the photons by the imperfection to be "elastic"). We will restrict our analysis to the particular case in which the Kraus operators defining the error channel are

$$\{E_k\} = \{\sqrt{p_1}\,\sigma_0^{\otimes 3}, \sqrt{p_2}\,\sigma_x^{\otimes 3}, \sqrt{p_3}\,\sigma_y^{\otimes 3}, \sqrt{p_4}\,\sigma_z^{\otimes 3}\},$$

where we assume $\sum p_i = 1$. Let us consider the simple three-qubit quantum circuit in Fig. 1.[3]

If we denote our ancillae as $|v\rangle$, $|u\rangle$, and our data qubit as $|\psi\rangle$ it can easily be shown[3] that

$$\left(\mathcal{U}_E^\dagger \circ \mathcal{E} \circ \mathcal{U}_E\right)(|v, u\rangle\langle v, u| \otimes |\psi\rangle\langle\psi|)$$

$$= \left(\sum_{i=0}^{3} p_i M_i |v, u\rangle\langle v, u| M_i^\dagger\right) \otimes |\psi\rangle\langle\psi|, \qquad (3)$$

where $\{M_i\} = \{\sigma_0^{\otimes 2}, \sigma_x^{\otimes 2}, -\sigma_x \otimes \sigma_y, \sigma_0 \otimes \sigma_z\}$. One qubit is thus protected against the fully correlated noise by encoding it with two ancillae, whose

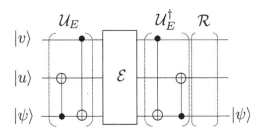

Fig. 1. Simplest NS, in terms of the number of CNOT gates, to avoid fully correlated noise defined by the errors $\{E_k\} = \{\sqrt{p_1}\,\sigma_0^{\otimes 3}, \sqrt{p_2}\,\sigma_x^{\otimes 3}, \sqrt{p_3}\,\sigma_y^{\otimes 3}, \sqrt{p_4}\,\sigma_z^{\otimes 3}\}, \sum p_i = 1$.

state can be quite arbitrary - in fact, it can be the uniformly mixed state. This scheme provides the *simplest* possible NS in terms of the number of CNOT gates under the present noise model. This statement is trivial when we consider a system of three (or more) qubits, because at least two CNOT gates are needed to correlate them. The non-existence of a NS with a two-qubit system under our noise model was shown in Ref. 3.

Possible applications of this scheme include its use to allow communication between two parties (conventionally designated as Alice and Bob) without a shared reference frame (SRF), provided that their respective bases differ by a rotation given by one of σ_i's.[3] The allowed rotations are limited to the σ_i's because we are considering a generalized definition of NS.[1,2] A more general theory of quantum communication without a SRF wherein the reference frames of Alice and Bob may differ by arbitrary rotations on account of the original, more restrictive definition of NS being used[4–10] can be found in Ref. 11. Note that in the framework of conventional DFS/NS, Alice and Bob may be light-years apart and they do not have to communicate in advance before qubits are sent; in our case, however, they need to establish the directions of the $x-$, $y-$ and $z-$axes of the Bloch sphere in advance using classical communication.

4. Experiments

We implemented our algorithm experimentally using a three-qubit NMR quantum computer. We used a JEOL ECA-500 NMR spectrometer. As a three-spin molecule, we employed ^{13}C-labeled L-alanine (98% purity, Cambridge Isotope) solved in D_2O.

We performed three sets of experiments corresponding to three different error operators (equivalent to artificially setting the value of one of the p_i's to 1 and the others to 0). We observed the data qubit, and applied quantum process tomography[12] to each set of data. A geometrical illustration of the action of our scheme on the data qubit is given in Fig. 2. Entanglement fidelities $F_e(\sigma_0, \mathcal{M})$ are summarized in the table. The analysis of our results shows that the proposed scheme does indeed eliminate the effects of fully correlated noises. We hypothesize that fidelity loss may be mostly due to an inhomogeneous control field (H_1 in NMR terminology) of pulses.[13] We expect that better results may be obtained by employing composite pulses.[14]

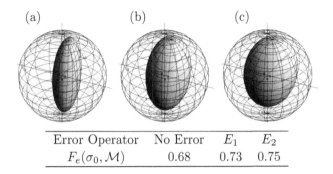

Error Operator	No Error	E_1	E_2
$F_e(\sigma_0, \mathcal{M})$	0.68	0.73	0.75

Fig. 2. Quantum process tomography results for three sets of experimental data corresponding to different error operators: (a) No Error, (b) E_1 (Fully Correlated X Error), and (c) E_2 (Fully Correlated Y Error), respectively. In each case, the outer Bloch sphere represents the set of all the pure (initial) states, while the inner surface represents the output states after the recovery process. The initial states are transformed to the final states by the map $\mathcal{M}$ defined by the encoding, error and decoding processes.

5. Conclusions

We demonstrate a simple quantum error correction scheme avoiding (restricted) fully correlated errors. There is no need for initialization of the ancillae, which can be in the uniformly mixed state. Experimental results show that our scheme eliminates the effects of fully correlated noise.

References

1. D. W. Kribs, R. Laflamme and D. Poulin, Phys. Rev. Lett. **94**, 180501 (2005); D. Kribs, R. Laflamme, D. Poulin, and M. Lesosky, Quantum Inf. Comput. **6**, 383 (2006).
2. M. A. Nielsen and D. Poulin, Phys. Rev. A **75**, 064304 (2007).
3. Y. Kondo, C. Bagnasco, and M. Nakahara, Phys. Rev. A **88**, 022314 (2013).
4. E. Knill, R. Laflamme and L. Viola, Phys. Rev. Lett. **84**, 2525 (2000).
5. S. De Filippo, Phys. Rev. A **62**, 052307 (2000).
6. C.-P. Yang and J. Gea-Banacloche, Phys. Rev. A **63**, 022311 (2001).
7. J. Kempe, D. Bacon, D. A. Lidar and K. B. Whaley, Phys. Rev. A **63**, 042307 (2001).
8. C.-K. Li, M. Nakahara, Y.-T. Poon, N.-S. Sze and H. Tomita, Phys. Rev. A **84**, 044301 (2011).
9. C.-K. Li, M. Nakahara, Y.-T. Poon, N.-S. Sze, H. Tomita, Phys. Lett. A **375**, 3255 (2011).
10. G. Chiribella, M. Dall'Arno, G.M. D'Ariano, C. Macchiavello, P. Perinotti, Phys. Rev. A **83**, 052305 (2011).
11. S. D. Bartlett, T. Rudolph, and R. W. Spekkens, Phys. Rev. Lett. **91**, 027901 (2003).

12. See, for example, Y. Kondo, J. Phys. Soc. Jpn. **76**, 104004 (2007), and references therein.
13. The control field inhomogeneity was considered in the following paper. E. H. Lapasar, K. Maruyama, D. Burgarth, T. Takui, Y. Kondo and M. Nakahara, New J. Phys. **14**, 013043 (2012).
14. M. H. Levitt, Prog. Nucl. Magn. Resonance Spectrosc. **18**, 61 (1986); T. Ichikawa, M. Bando, Y. Kondo and M. Nakahara, Phil. Trans. R. Soc. A **370**, 4671 (2012), and references therein.

BLACK HOLE PREDICTABILITY, CLASSICAL AND QUANTUM

AKIHIRO ISHIBASHI

Department of Physics, Kinki University,
Higashi-Osaka, Osaka 577-8502, Japan
E-mail: akihiro@phys.kindai.ac.jp

This article attempts to give a brief overview of recent approaches to the black hole information loss puzzle by exploiting the gauge-gravity duality.

1. Introduction

"The black holes of nature are the most perfect macroscopic objects there are in the universe: the only elements in their construction are our concepts of space and time. And since the general theory of relativity provides only a single unique family of solutions for their descriptions, they are the simplest objects as well." describes Chandrasekhar.[1] As the *simplest objects*, black holes have not only been one of the most interesting astrophysical objects but also have played a role of theoretical laboratory, providing us deep insights into the global structure of spacetime and quantum aspects of gravity.

The *single unique family* mentioned above is the Kerr metric, which is a solution to the vacuum Einstein equations with only two parameters, the mass and angular momentum. The mathematical basis of this statement is the black hole *uniqueness* or *no hair* theorem, which states that an asymptotically flat, stationary black hole solution to the Einstein equations with, at most Maxwell field, can be uniquely specified by the three conserved charges, i.e., the mass M, the angular momentum J, and the electric charge Q [see, e.g., reviews[23] and references therein]. Thus, a tremendous number of black holes in our universe can be accurately described by the Kerr metric.

Since astrophysical black holes are the end product of gravitational collapse, a consequence of the no hair theorem is that all information about the collapsing body, such as multipole moments, is apparently lost from the outside region except the three *hair* M, J, and Q. As far as classical theory is considered, this is not a problem since it is just that all the information is hidden inside the event horizon. However, taking into account quantum effects drastically changes a possible consequence, raising a profound question concerning the predictability. In this article we shall briefly review the nature of this puzzle and summarize recent proposals as a possible resolution to the puzzle.

2. Black hole thermodynamics

We begin with briefly recapitulating thermodynamic aspects of black holes. The claim of the no hair theorem that a stationary black hole is uniquely determined by a small number of parameters has an immediate resemblance to the fact that an equilibrium thermodynamic system is characterized by a small number of state parameters.[4] Immediately after Hawking[5] showed the area theorem which states that under certain physical conditions, the surface area A of the horizon cross-section can never decrease with time, Bekenstein[6] has proposed that the area of the black hole horizon is the black hole entropy, in accord with 2nd-law of ordinary thermodynamics that the entropy S never decrease. In 1973, Bardeen, Carter, and Hawking[7] have derived the 1st-law of black hole mechanics, $\kappa \delta A = 8\pi G(\delta M - \Omega_H \delta J)$, identifying that the horizon area A corresponds to the entropy S, the surface gravity κ to the temperature T, the mass M to the energy E of an equilibrium thermodynamic system, and the horizon angular velocity Ω_H times the angular momentum J to work term. However, since within the classical framework the temperature of a black hole is absolute zero, these results could just be a mathematical resemblance. In 1975, by considering quantum effects on the background black hole geometry, Hawking[8] showed that a black hole actually can emit thermal radiation with the temperature specified by the surface gravity. The discovery of this perfect black body radiation, or *Hawking radiation*, has established the correspondence between the black hole mechanics and thermodynamics, identifying in particular that the black hole entropy is the quarter of the horizon area.

3. Breakdown of predictability

According to the black hole thermodynamics, a black hole actually carries an enormous amount of the entropy proportional to its horizon cross-section area $S \sim A/4 \sim M^2$. Also a black hole can emit black body radiation with the temperature $T \sim \kappa \sim 1/M$. Therefore the smaller a black hole is, the higher its Hawking temperature becomes. This implies that once the backreaction of the Hawking radiation is taken into account, the black hole starts evaporating, loosing its mass, and becomes hotter and hotter and eventually explodes. The Hawking radiation is a perfect black body radiation and therefore cannot carry any information from inside the black hole. Then one can ask whether all the information hidden inside the event horizon will be lost completely or not. In terms of quantum theory, initial data that collapse to form a black hole with total mass M and angular momentum J can be described as a pure state, Ψ. The black hole emits Hawking radiation and evaporates, leaving thermal radiations in the future region with no black hole, which can be characterized by mixed states $\{p_n, \Psi_n\}$ with a density matrix $\rho = \sum_n p_n \Psi_n^* \Psi_n$. This means that the whole process of a black hole formation and evaporation is a time evolution from the pure state Ψ to the mixed state $\{p_n, \Psi_n\}$, which cannot be a unitary evolution. Therefore a black hole evaporation by Hawking radiation implies a breakdown of quantum mechanical predictability. The latter is also known as the black hole information loss puzzle.[9] The process of a black hole formation and evaporation is sometime represented by the spacetime diagram given by Kodama[10] (see Figure 1).

4. Resolution to the information puzzle

If the information is really lost, quantum mechanics needs to be modified. If not, then general relativistic picture of black hole spacetime needs be modified. A possible resolution to this information puzzle recently proposed has exploited the gauge-gravity duality, which claims that quantum gravity is equivalent to a non-gravitational gauge theory living at spacetime infinity. This duality is often called *holographic* in the sense that the gauge theory dual to the bulk spacetime gravity is living on a lower dimensional spacetime at the spacetime infinity. One of the most impressive and successful examples of this duality is given by gravity in Anti-de Sitter (AdS) spacetime and conformal field theory (CFT) at AdS infinity,[11] which is often referred to as the *AdS/CFT correspondence*. The AdS_D is a Lorentzian version of a D-dimensional hyperboloid, whose boundary at infinity is a lower, $(D-1)$-dimensional spacetime $\mathrm{R} \times S^{D-2}$. A concrete formulation[12][13] of the correspondence is that the partition function of a boundary gauge field is equivalent to that of the bulk AdS gravity, $Z_{\text{gauge}} = Z_{\text{AdS}}$. When the bulk AdS spacetime contains a black hole, the corresponding boundary theory is a thermal gauge field theory. Now let us assume that the AdS/CFT correspondence is correct. Then a black hole formation and evaporation in the

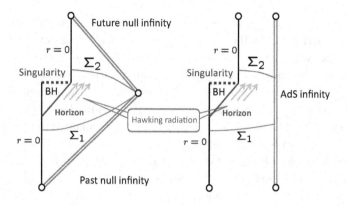

Fig. 1. A spacetime diagram for a black hole formation and evaporation. As far as the present author is aware, this type of spacetime picture for the black hole evaporation was first proposed by Kodama.[10] The left is for the asymptotically flat case, while the right for the asymptotically AdS case. On the initial hypersurface $\Sigma_1 \approx \mathbf{R}^3$, one prepares initial data for gravitational collapse to a black hole in terms of a pure state Ψ. On $\Sigma_2 \approx \mathbf{R}^3$, one observes the Hawking radiation described by a density matrix $\rho = \sum_n p_n \Psi_n^* \Psi_n$. The evolution from Ψ to $\{p_n, \Psi_n\}$ is not unitary. It should be noticed that neither Σ_1 nor Σ_2 is a Cauchy surface.

AdS spacetime should be described in terms of the boundary CFT, which obeys ordinary quantum mechanics, hence unitary evolution. As far as we are looking at the spacetime diagram depicted in Figure 1, it would be hard to imagine how to reconcile the boundary unitary evolution and apparently non-unitary looking evolution in the bulk spacetime. In fact, both the initial hypersurface Σ_1 and a late time hypersurface Σ_2 are *not* Cauchy surfaces: Cauchy developments of Σ_1 and Σ_2 cannot cover each other. In order to reconcile the unitarily evolving picture of the boundary quantum theory with time evolution in the bulk spacetime, one needs to discard the idea that a spacetime can be described by a single spacetime diagram as in Figure 1. Instead one has to view that the relevant spacetime picture is given, as a consequence of quantum gravity, by *superposition* of all possible spacetime geometries that satisfy the same asymptotic boundary condition at AdS infinity.[14] Once this spacetime picture for the black hole formation and evaporation is adopted, then the information should not be lost, and quantum mechanics is predictable.

5. Summary

Black holes have provided profound questions such as the origins of its entropy, quantum evaporation, the information loss puzzle. The idea of gauge-gravity duality offers a possible resolution to the information puzzle, but gives rise to further open questions;[15] e.g, how does the information come out of the black hole? The holographic nature of the suggested resolution indicates one may need to violate locality.

Acknowledgments

AI is supported by the Grant-in-Aid for Scientific Research Fund of the JSPS (C)No. 22540299.

References

1. S. Chandrasekhar, *The Mathematical Theory of Black Holes*, (Oxford University Press, Oxford, 1983)
2. P. T. Chrusciel, J. L. Costa, and M. Heusler, *Stationary Black Holes: Uniqueness and Beyond*, Living Reviews in Relativity **15**, 7 (2012)
3. S. Hollands and A. Ishibashi, *Black hole uniqueness theorems in higher dimensional spacetimes* Class. Quant. Grav. **29**, 163001 (2012)
4. R. M. Wald, *The Thermodynamics of Black Holes*, Living Reviews in Relativity **4**, 2001-6 (2001)
5. S. W. Hawking, *Black holes in general relativity*, Commun. Math. Phys. **25**, 152-166 (1972)
6. J. D. Bekenstein, *Black holes and entropy*, Phys. Rev. D **7**, 2333-2346 (1973)
7. J. M. Bardeen, B. Carter, and S. W. Hawking, *The Four laws of black hole mechanics*, Commun. Math. Phys. **31**, 161-170 (1973)
8. S. W. Hawking, *Particle Creation by Black Holes* Commun. Math. Phys. **43**, 199 (1975)
9. S. W. Hawking, *Breakdown of predictability in gravitational collapse*, Phys. Rev. D **14**, 2460 (1976)
10. H. Kodama, *Conserved Energy Flux for the Spherically Symmetric System and the Backreaction Problem in the Black Hole Evaporation*, Prog. Theor. Phys. **63**, 1217 (1980)
11. J. M. Maldacena, *The Large N Limit of Superconformal Field Theories and Supergravity*, Adv. Theor. Math. Phys. **2**, 231 (1998)
12. S. S. Gubser, I. R. Klebanov, and A. M. Polyakov, *Gauge theory correlators from noncritical string theory*, Phys. Lett. B **428**, 105 (1998)
13. E. Witten, *Anti-de Sitter space and holography*, Adv. Theor. Math. Phys. **2**, 253 (1998)
14. S. W. Hawking, *Information loss in black holes*, Phys. Rev. D **72**, 084013 (2005)
15. A. Almheiri, D. Marolf, J. Polchinski, and J. Sully, *Black Holes: Complementarity or Firewalls*, JHEP **1302**, 062 (2013)

CLASSICAL FIELD SIMULATION OF FINITE-TEMPERATURE BOSE GASES

TOSHIHIRO SATO

Computational Condensed Matter Laboratory, RIKEN,
Wako, Saitama 351-0198, Japan
E-mail: toshihiro.sato@riken.jp

We study the finite-temperature behavior of a dilute Bose gas using projected Gross-Pitaevskii equation by comparing to that of quantum Monte Carlo calculation in a two dimensional homogeneous system. By investigating the behavior of the correlation function, obtained results indicate that projected Gross-Pitaevskii equation is effective model of the Bose atoms with the long wavelength at low temperatures including the superfluid transition temperature. As the demonstration in a two dimensional homogeneous system, we find that this model describe the characteristic of vortex excitation indicated the superfluid transition.

Keywords: Quantum gases, Superfluid

1. Introduction

Since achievement of Bose-Einstein condensation (BEC) in dilute gases,[1–4] many phenomena of a quantum Bose field have been actively studied from theoretical and experimental aspects. As for recent works in dilute Bose gases, it is interesting to investigate how dynamical characteristics and the phase transition appear. The properties of Bose condensates near zero-temperature, where most of the atoms are in the condensate, are well-described by the Gross-Pitaevskii equation (GPE).[5–9] Recently, it has been proposed that the GPE is also applicable to represent Bose gases at a finite temperature, and is demonstrated for several finite-temperature behaviors by several groups.[10–18] The basic concept is that Bose atoms with the long wavelength are well-described by a classical field as is well known in the case of electromagnetic fields. At low temperatures, many Bose atoms are occupied near the lowest energy modes with the long wavelength and the GPE represented by a classical field is useful model. However, higher-temperature Bose atoms are occupied over the higher energy modes, which are not approximated as the classical field. In recent works, Davis *et al.* have been performed the improvement of this problem and proposed the effective model of finite-temperature Bose gases. The idea of this model is to introduce the energy cutoff through the use of a projector that is diagonal for the Hamiltonian of the single-particle basis of the system. The energy cutoff separates a quantum Bose field of the system at finite-temperature between the regions of lower energy modes as described by classical field (classical region) and other regions of sparsely occupied higher energy modes (incoherent region). Ignoring the collision process of Bose atoms between the classical modes and the incoherent modes because the energy and particle fluctuation of the process between two regions are very small at equilibrium, the equation of motion for the classical region is termed 'projected Gross-Pitaevskii equation (PGPE).[19] The first calculations using PGPE are demonstrated for the finite-temperature behaviors of a three-dimensional (3D) dilute Bose gas .[20–22] Some experimental comparative studies are performed for the BEC transition temperature of dilute Bose gases trapped in the 3D harmonic potential[23,24] and the superfluidity in the quasi-two-dimensional (2D) harmonic potential.[25–30] While the obtained results are confirmed qualitatively, it is difficult to catch how PGPE describes the finite-temperature behavior of a dilute Bose gas quantitatively.

In this paper, we study the finite-temperature behavior of a dilute Bose gas using PGPE by comparing to that of quantum Monte Carlo (QMC) method.[31] We focus on a 2D homogeneous Bose gas and investigate the behavior of the correlation function quantitatively.

2. Model and Method

PGPE in a 2D homogeneous Bose gas is described by

$$i\hbar\frac{\partial\psi_i(t)}{\partial t} = -\frac{t}{Z}\sum_j[\psi_j(t) - \psi_i(t)] + U\hat{P}\{|\psi_i(t)|^2\psi_i(t)\}, \tag{1}$$

where t is the hopping matrix elements, U is the repulsive interaction and Z is coordination number and we set $Z = 4$ in following calculation. $\psi_i(t)$ denotes the field in the classical region C and is given by

$$\psi_i(t) = \sum_{\mathbf{k}\in C} c_{\mathbf{k}}(t)\frac{e^{i\mathbf{k}\cdot\mathbf{r}_i}}{\sqrt{V}}, \tag{2}$$

where V is the system volume. The classical region C is defined as the subspace that consists of low-energy modes below the energy cutoff $k_{\mathrm{cut}}(T)$. $k_{\mathrm{cut}}(T)$ is the energy level at which the mean occupation number of a

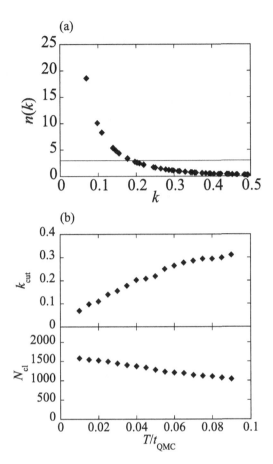

Fig. 1. (a) Wave number dependence of the momentum distribution $n(k)$ at $T/t_{\mathrm{QMC}} =$ 0.04. Solid line is the mean occupation number $n_{\mathrm{cut}} = 3$. (b) Temperature dependence of $k_{\mathrm{cut}}(T)$ and $N_{\mathrm{cl}}(T)$ for $N_{\mathrm{tot}} = 0.1 \times V$. Temperature T/t_{QMC} is used in QMC calculations.

single-particle state takes a value $n_{\mathrm{cut}} = 3$.[32] The high-energy states are eliminated by the operator $\hat{P}\{F(\mathbf{r})\} = \sum_{n \in \mathcal{C}} \phi_n(\mathbf{r}) \int d\mathbf{r}' \phi_n^*(\mathbf{r}') F(\mathbf{r}')$, where $\phi_n(\mathbf{r})$ is a single-particle eigenstate of the non-interacting model. In our calculations, we set $V = 128 \times 128$, total particle $N_{\mathrm{tot}} = 0.1 \times V$ and $U/t = 0.1$. Following previous studies on the PGPE calculations,[32] we also estimate $k_{\mathrm{cut}}(T)$, the number of particles $N_{\mathrm{cl}}(T)$ in the classical region C at a fixed N_{tot}. To evaluate the results solved by Eq. (1), we perform the QMC computations. In the QMC computations, we treat the 2D Bose-Hubbard

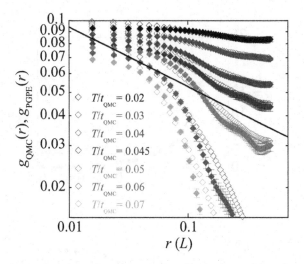

Fig. 2. Temperature dependence of the correlation function $g(r)$. ♦ is the results of PGPE calculation and ◊ is those of QMC calculation. The solid-line denotes $g(r) = r^{-1/4}$. L is the length of system.

model described by

$$H = -\frac{t}{Z} \sum_{\langle i,J \rangle} (b_i^\dagger b_j + b_i b_j^\dagger) + \frac{U}{2} \sum_i b_i^\dagger b_i^\dagger b_i b_i, \qquad (3)$$

where b_i ($b_i^\dagger$) is creation (annihilation) operators.

We perform the comparison results of PGPE with that of QMC calculation as follows. We first estimate the temperature dependence of the momentum distribution $n(k)$ by QMC calculations in Eq. (3). The results $n(k)$, $N_{cl}(T)$ and initial configuration $c_k(0) = \sqrt{n(k)}$ to PGPE calculations are given by using $k_{cut}(T)$ associated with $n_{cut} = 3$. Figures 1 present the results of $n(k)$ at $T/t_{QMC} = 0.04$, the temperature dependence of $k_{cut}(T)$ and $N_{cl}(T)$. Second, we make the initial state $\psi_i(0)$ from $c_k(0) = \sqrt{n(k)}$ with a randomized phase, and employ it as a initial condition of PGPE. From these results, we perform the PGPC calculation and investigate the behavior of the correlation function (QMC method : $g_{QMC}(r) = 1/V \sum_i \langle b_i^\dagger b_{i+r} \rangle$, PGPE : $g_{PGPE}(r) = 1/V \sum_i \langle \psi_i^* \psi_{i+r} \rangle$).

3. Results

In 2D homogeneous bosonic systems, it is well known that the finite temperature transition to the superfluid state obeys the Kosterlitz-Thouless

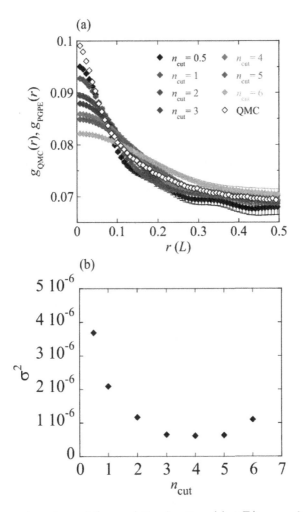

Fig. 3. (a) n_{cut} dependence of the correlation function $g(r)$ at $T/t_{QMC} = 0.03$. ♦ is the results of PGPE calculation and ◊ is those of QMC calculation. L is the length of system. (b) n_{cut} dependence of the standard deviation σ for $r \geq 1/k_{cut}$ at $T/t_{QMC} = 0.03$.

(KT) type and is characterized by binding/unbinding of vortex pairs.[33–35] For 2D superfluid state, the correlation function shows the power-law decay $g(r) \sim r^{-\eta}$ below the transition temperature T_{KT} and the effective exponent η takes a value $1/4$ at T_{KT}. In Figure 2, we plot the temperature dependence of the correlation function $g_{QMC}(r)$ and $g_{PGPE}(r)$. From the obtained results of QMC calculations, the power-law decay with distance is observed for $T/t_{QMC} \leq T_{KT}/t_{QMC} = 0.045$. On the other hand, the

Fig. 4. Instantaneous density profiles $n(x, y)$ (upper) and phase profiles $\theta(x, y)$ (lower) solving PGPE of Eq. (1). (a) $T/t_{\mathrm{QMC}} = 0.03$ and (b) $T/t_{\mathrm{QMC}} = 0.05$. L is the length of the system. In the phase profiles, vortex (+) and antivortex (-) are plotted.

well-developed thermal fluctuation for $T/t_{\mathrm{QMC}} > T_{\mathrm{KT}}/t_{\mathrm{QMC}}$ makes $g(r)$ exponentially decays as expected. The obtained results of PGPE calculations clearly indicate the argument of those of QMC calculations below $T/t_{\mathrm{QMC}} \leq 0.05$.

The classical region described by PGPE depends on the energy cutoff k_{cut} detected by the mean occupation number n_{cut} as can use the classical field description, based on the electromagnetic fields. We also investigate n_{cut} dependence at temperature, which clear agrees with the results of QMC calculation. In Figure 3 (a), we shows the n_{cut} dependence of $g_{\mathrm{PGPE}}(r)$ at $T/t_{\mathrm{QMC}} = 0.03$. PGPE is the effective model to the regions of the long wave length (low energy) below k_{cut}, and it is importance whether agrees with the behavior of $g(r)$ for $r \geq 1/k_{\mathrm{cut}}$. Therefore, we estimate the standard deviation σ of $g_{\mathrm{PGPE}}(r)$ from QMC calculation $g_{\mathrm{QMC}}(r)$ for $r \geq 1/k_{\mathrm{cut}}$,

$$\sigma^2 = \frac{1}{L/2 - 1/k_{\mathrm{cut}}} \int_{1/k_{\mathrm{cut}}}^{L/2} dr [g_{\mathrm{PGPE}}(r) - g_{\mathrm{QMC}}(r)]^2, \qquad (4)$$

where L is the system length. In Figure 3 (b), we plot the n_{cut} dependence of σ at $T/t_{\text{QMC}} = 0.03$. The obtained results indicate that σ is minimized for $3 \leq n_{\text{cut}} \leq 5$ because σ gradually increases for $n_{\text{cut}} \leq 2$ and $5 \leq n_{\text{cut}}$. The former results cause that PGPE description includes the region as not can use the classical field description. On the other hands, the latter disagreement expects that the too small classical region to describe the finite-temperature Bose gases. These more detailed investigation is in progress and is published elsewhere.[36]

In the PGPE calculations, one can catch the phase distribution directly and simulate the dynamics of a quantum field. In Figure 4, we show the results of PGPE calculation for the instantaneous density profiles $n(x, y)$ and phase profiles $\theta(x, y)$ at $T/t_{\text{QMC}} = 0.03$ and $T/t_{\text{QMC}} = 0.05$. At $T/t_{\text{QMC}} = 0.03$ below $T_{\text{KT}}/t_{\text{QMC}}$, it is obvious from the phase profiles that the system is superfluidity with the phase coherent. At a high temperature $T/t_{\text{QMC}} = 0.05$, many vortices and antivortices are observed and the system indicates the normal state. These obtained results particularly describe the characteristic of vortex excitation indicated the superfluid transition in 2D homogeneous systems.

4. Summary

In summary, we have studied the finite-temperature behavior of a dilute Bose gas using projected Gross-Pitaevskii equation by comparing to that of quantum Monte Carlo calculation in a 2D homogeneous system. By investigating the behavior of the correlation function, we confirm that the behaviors of long range correlation at the low temperature including the superfluid transition temperature are well-described. Moreover, we find that the applicable values of n_{cut} exist to use PGPE. PGPE can perform the large-scale calculations to the experimental comparative studies on a dilute Bose gas and will be the powerful tool for the dynamics of a finite-temperature dilute Bose gas.

Acknowledgment

The author is grateful to Yasuyuki Kato, Takafumi Suzuki, and Naoki Kawashima for collaborations in this work. The present work is financially supported by MEXT Grant-in-Aid for Young Scientists (B) (21740245), MEXT Grant-in-Aid for Scientific Research (B) (19340109), MEXT Grant-in-Aid for Scientific Research on Priority Areas "Novel States of Matter Induced by Frustration" (19052004), Next Generation Supercomputing

Project, Nanoscience Program, MEXT, Japan, and Global COE Program "the Physical Sciences Frontier", MEXT, Japan. The computation in the present work was performed on computers at the Supercomputer Center, Institute for Solid State Physics, University of Tokyo.

References

1. K. B. Davis, M-O. Mewes, M. R. Andrews, N. J. van Druten, D. S. Durfee, D. M. Kurn, and W. Ketterle, *Phys. Rev. Lett.* **75**, 3969 (1995).
2. M. Anderson, J. R. Ensher, M. R. Matthews, C. E. Wieman, and E. A. Cornell, *Science* **269**, 198 (1995).
3. C. C. Bradley, C. A. Sackett, J. J. Tollet, and R. Hulet, *Phys. Rev. Lett.* **75**, 1687 (1995).
4. C. A. Sackett, J. J. Tollet, and R. Hulet, *Phys. Rev. Lett.* **79**, 1170 (1997).
5. A. Röhrl, M. Naraschewski, A. Schenzle, and H. Wallis, *Phys. Rev. Lett.* **78**, 4143 (1997).
6. M. Tsubota, K. Kasamatsu, and M. Ueda, *Phys. Rev. A* **65**, 023603 (2002).
7. K. Kasamatsu, M. Tsubota, and M. Ueda, *Phys. Rev. Lett.* **91**, 150406 (2003).
8. N. Sasa, M. Machida, and H. Matsumoto, *J. Low Temp. Phys.* **138**, 617 (2005)
9. T. Sato, T. Ishiyama and T. Nikuni, *Phys. Rev. A* **76**, 053628 (2007).
10. B. V. Svistunov, *J. Moscow Phys. Soc.* **1**, 373 (1991).
11. Yu. Kagan, B. V. Svistunov, and G. V. Shlyapnikov, *Zh. Eksp. Teor. Fiz.* **101**, 528 (1992) (Engl. transl. *Sov. Phys. -JETP*, **75**, 387 (1992)).
12. Yu. Kagan and B. V. Svistunov, *Zh. Eksp. Teor. Fiz.* **105**, 353 (1994) (Engl. transl. *Sov. Phys. -JETP*, **78**, 187 (1994)).
13. Yu. Kagan and B. V. Svistunov, *Phys. Rev. Lett.* **79**, 3331 (1997).
14. K. Damle, S. M. Majundar, and S. Sachdev, *Phys. Rev. A* **54**, 5037 (1996).
15. R. J. Marshall, G. New, K. Burnett, and S. Choi, *Phys. Rev. A* **59**, 2085 (1999).
16. H. T. C. Stoof and M. Bijlsma, *J. Low Temp. Phys.* **124**, 431 (2001).
17. A. Sinatra, C. Lobo, and Y. Castin, *Phys. Rev. Lett.* **87**, 210404 (2001).
18. K. Gòral, M. Gajda, and K. Rzażewski, *Opt. Express* **8**, 92 (2001).
19. M. J. Davis, R. J. Ballagh, and K. Burnett, *J. Phys. B* **34**, 4487 (2001).
20. M. J. Davis, S. A. Morgan, and K. Burnett, *Phys. Rev. Lett.* **87**, 160402 (2001).
21. M. J. Davis, S. A. Morgan, and K. Burnett, *Phys. Rev. A* **66**, 053618 (2002).
22. P. B. Blakie and M. J. Davis, *Phys. Rev. A* **72**, 063608 (2005).
23. F. Gerbier, J. H. Thywissen, S. Richard, M. Hugbart, P. Bouyer, and A. Aspect, *Phys. Rev. Lett.* **92**, 030405 (2004).
24. M. J. Davis and P. B. Blakie, *Phys. Rev. Lett.* **96**, 060404 (2006).
25. S. Stock, Z. Hadzibabic, B. Battelier, M. Cheneau, and J. Dalibard, *Phys. Rev. Lett.* **95**, 190403 (2005).
26. Z. Hadzibabic, P. Krüger, M. Cheneau, B. Battelier, and J. Dalibard, *Nature* **441**, 1118 (2006).

27. P. Krüger, Z. Hadzibabic, and J. Dalibard, *Phys. Rev. Lett.* **99**, 040402 (2007).

28. T. P. Simula and P. B. Blakie, *Phys. Rev. Lett.* **96**, 020404 (2006).

29. T. P. Simula, M. J. Davis, and P. B. Blakie, *Phys. Rev. A* **77**, 023618 (2008).

30. T. Sato, T. Suzuki, and N. Kawashima, *Phys. Rev. A* **61**, 025601 (2010).

31. Y. Kato, T. Suzuki, and N. Kawashima, *Phys. Rev. E* **75**, 066703 (2007).

32. P. B. Blakie and M. J. Davis, *J. Phys. B* **40**, 2043 (2007).

33. N. D. Mermin and H. Wagner, *Phys. Rev. Lett.* **17**, 1133 (1966).

34. P. C. Hohenberg, *Phys. Rev.* **158**, 383 (1967).

35. V. L. Berezinskii, *Sov. Phys. JETP* **32**, 493 (1971) ; **34**, 610 (1972).

36. J. M. Kosterlitz and D. J. Thouless, *J. Phys. C* **6**, 1181 (1973).

37. J. M. Kosterlitz, *J. Phys. C* **7**, 1046 (1974).

38. T. Sato, Y. Kato, T. Suzuki, and N. Kawashima, *Phys. Rev. E* **85**, 050105(R) (2012).

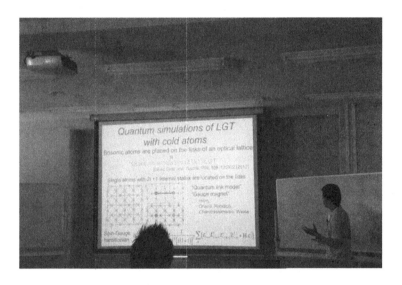

ATOMIC QUANTUM SIMULATIONS OF LATTICE GAUGE THEORY: EFFECT OF GAUGE SYMMETRY BREAKING

KENICHI KASAMATSU[*,1], IKUO ICHINOSE[2] and TETSUO MATSUI[1]

[1] *Department of Physics, Kinki University, Higashi Osaka, Osaka, 577-8502, Japan*
[2] *Department of Applied Physics, Nagoya Institute of Technology, Nagoya, 466-8555, Japan*
E-mail: kenichi@phys.kindai.ac.jp

We investigate the effect of gauge symmetry breaking in an atomic quantum simulator of U(1) lattice gauge theory, finding that a system of cold atoms in an optical lattice with a general set of parameters can be implemented as a simulator of "U(1) lattice gauge-Higgs" model with asymmetric nearest-neighbor Higgs coupling. The gauge-Higgs coupling in the imaginary time direction naturally arises from the violation of the U(1) local gauge invariance of the simulators caused by the deviation from the fine-tuned system parameters. A general method to supply the Higgs coupling in all space-time directions may be realized by coupling atoms in an optical lattice to another particle reservoir filled with the Bose-condensed atoms via laser-induced transitions. Clarification of the dynamics of this gauge-Higgs model sheds some lights upon various unsolved problems including the inflation process of the early universe.

Keywords: Quantum simulation; Lattice gauge theory; Bose-Einstein condensation; Optical lattice; Gauge-Higgs model

1. Introduction

Recently, there are strong attentions for the possibility of using ultracold atoms in an optical lattice (OL) as a quantum simulator for lattice gauge theories (LGTs)[1-3] . The basic ingredients are to place several kinds of cold atoms on the "links" between the sites of an OL according to certain rules.[4-10] Several papers gave proposals for pure U(1) LGTs in Refs.[4-6] and later extended to quantum electrodynamics with dynamical fermionic matter[7,8] and non-Abelian gauge models[9,10] .

Shortly after their introduction by Wilson[11] , LGTs have been studied quite extensively, mainly in high-energy physics, by using both analytical methods and Monte Carlo simulations, and their various properties have now been clarified. However, the above mentioned approach using cold atoms in an OL provides us with another interesting method for studying LGTs. The atomic quantum simulations allow one to address problems which cannot be solved by conventional Monte Carlo methods because of sign problem. One drawback of the atomic simulators is that the equivalence to the gauge system is established only under some specific conditions. For example, in Refs.[4–10] , one needs to fine-tune a set of interaction parameters to realize the local gauge symmetry which is essential to a quantum simulator of LGTs. In other words, the local gauge symmetry is explicitly lost when these parameters deviate from their optimal values.

The above-mentioned point naturally poses us serious and important questions on the stability of gauge symmetry, and potential subtlety of experimental results of cold atoms as simulators of LGTs, because the above conditions are generally not satisfied exactly or easily violated in actual systems. We address this problem semi-quantitatively and exhibit the allowed range of violation of the above conditions, such as the regime of interaction parameters, within which the results can be regarded as having LGT properties.[12] In addition, we propose that the cold atomic system in question may be used as a quantum simulator of a wide class of "U(1) gauge-Higgs model", i.e., a Ginzburg-Landau-type model in the London limit coupled with the gauge field.[12] All of the previous literatures[4–10] involve no interactions between gauge and Higgs fields. Revealing the dynamics of the gauge-Higgs model should offer us important insights on several fields including inflational cosmology[13] , high-energy physics, and condensed matter theory (superconductivity,[14] etc.).

2. U(1) pure lattice gauge theory

We first briefly review the compact U(1) pure LGT as the reference system of the present study. The word "pure" implies that the system contains only gauge fields and no other fields such as quarks, etc. The system is still nontrivial because the gauge field on lattice is self-interacting.

The path-integral representation of the partition function Z is given by

$$Z = \int [dU] \exp(A), \quad \int [dU] \equiv \prod_{x,\mu} \int_0^{2\pi} \frac{d\theta_{x\mu}}{2\pi},$$

$$A = \frac{c_2}{2} \sum_x \sum_{\mu < \nu} \bar{U}_{x\nu} \bar{U}_{x+\nu,\mu} U_{x+\mu,\nu} U_{x\mu} + \text{c.c.} = c_2 \sum_x \sum_{\mu < \nu} \cos \theta_{x\mu\nu}, \quad (1)$$

where $\theta_{x\mu\nu} \equiv \nabla_\mu \theta_{x\nu} - \nabla_\nu \theta_{x\mu}$, $U_{x\mu} \equiv e^{i\theta_{x\mu}}$, and $\nabla_\mu f_x \equiv f_{x+\mu} - f_x$. The subscript $x = (x_1, x_2, x_3, x_4)$ is the site index of the 3+1=4D lattice (x_4 is the imaginary time in the path-integral approach) and μ and ν ($= 1, 2, 3, 4$) are the direction indices that we also use as the unit vectors in the μ and ν-th directions. The angle variable $\theta_{x\mu} \in [0, 2\pi)$ and its exponential $U_{x\mu}$ are the gauge variables defined on the link $(x, x+\mu)$. The bar in $\bar{U}_{x\mu}$ implies complex conjugate, and c_2 ($\equiv 1/e^2$) is the inverse self-gauge-coupling constant. The product of four $U_{x\mu}$ is invariant under the local (x-dependent) U(1) gauge transformation,

$$U_{x\mu} \to U'_{x\mu} \equiv V_{x+\mu} U_{x\mu} \bar{V}_x, \qquad V_x \equiv e^{i\Lambda_x}, \tag{2}$$

and so are the field strength $\theta_{x\mu\nu}$ and the action A.

The variable $\theta_{x\mu}$ is related to the vector potential $A_\mu(x)$ in the continuum space-time as $\theta_{x\mu} = aeA_\mu(x)$ where a is the lattice spacing. In the formal continuum limit $a \to 0$, the action is reduced to $A \to -(1/4) \int d^4x F_{\mu\nu}(x) F_{\mu\nu}(x)$ with $F_{\mu\nu} \equiv \partial_\mu A_\nu - \partial_\nu A_\mu$ as it should be. The partition function Z can be defined without a gauge fixing due to the compactness $\int [dU]1 = 1$ in contrast with $\int_{-\infty}^{\infty} dA_\mu(x) = \infty$. It is known that the system Eq. (1) has a weak first-order phase transition at $c_2 = c_{2c} \simeq 1.0$.[15] For $c_2 < c_{2c}$ ($> c_{2c}$) the system is in the confinement (Coulomb) phase in which the fluctuations of $\theta_{x\mu}$ are strong (weak). In the Coulomb phase, $\theta_{x\mu}$ describes almost-free massless particles, which correspond to photons in electromagnetism.

To obtain the quantum Hamiltonian $\hat{H}$ for Z, let us focus on the space-time plaquette term $\cos\theta_{xi4}$ in Z with the spatial direction index i ($= 1, 2, 3$) and rewrite it as

$$\exp(c_2 \cos\theta_{xi4}) \simeq \sum_{m_{xi} \in \mathbf{Z}} \exp\left[-\frac{c_2}{2}(\theta_{xi4} - 2\pi m_{xi})^2\right]$$

$$\propto \sum_{E_{xi} \in \mathbf{Z}} \exp\left[-iE_{xi}(\nabla_i\theta_{x4} - \nabla_4\theta_{xi}) - \frac{1}{2c_2}E_{xi}^2\right], \tag{3}$$

where we used the Villain (periodic Gaussian) approximation in the first line and Poisson's summation formula in the second line. The term $iE_{xi}\nabla_4\theta_{xi} \simeq id\tau E_{xi}\dot{\theta}_{xi}$ (τ is the imaginary time and $\dot{f} \equiv df/d\tau$) shows that the integer-valued field E_{xi} on the spatial link $(x, x+i)$ is the conjugate momentum of θ_{xi}. Thus, the corresponding operators at spatial site $r = (x_1, x_2, x_3)$ satisfy the canonical commutation relation $[\hat{E}_{ri}, \hat{\theta}_{r'i'}] = -i\delta_{rr'}\delta_{ii'}$. The operator $\hat{E}_{ri}$ represents the electric field in electromagnetism but has integer eigenvalues owing to the compactness (periodicity) of A under $\theta_{x\mu} \to \theta_{x\mu} + 2\pi$.

The integration over θ_{x4} can be done as

$$G \equiv \int \prod_x d\theta_{x4} \exp(-i \sum_{x,i} E_{xi} \nabla_i \theta_{x4}) = \prod_x \delta_{Q_x,0},$$

$$Q_x \equiv \sum_i \nabla_i E_{xi}, \tag{4}$$

where we used $\sum_{x,i} E_{xi} \nabla_i \theta_{x4} = -\sum_{x,i} \nabla_i E_{xi} \cdot \theta_{x4}$, which holds for a lattice with periodic boundary conditions.

One may check that the quantum Hamiltonian $\hat{H}$ which gives Z at the inverse temperature β is just the one known as the Kogut-Susskind (KS) Hamiltonian,[16]

$$\hat{H} = \frac{1}{2c_2\Delta\tau} \sum_{r,i} \hat{E}_{ri}^2 - \frac{c_2}{\Delta\tau} \sum_{r,i<j} \cos\hat{\theta}_{rij} \tag{5}$$

with $\Delta\tau$ ($\equiv \beta/N$) being the short-time interval in the τ direction. By inserting the complete sets $\hat{1}_E = \prod_{r,i} \sum_{E_{ri}} |\{E_{ri}\}\rangle\langle\{E_{ri}\}|$ and $\hat{1}_\theta = \prod_{r,i} \int d\theta_{ri} |\{\theta_{ri}\}\rangle\langle\{\theta_{ri}\}|$ in between the short-time Boltzmann factors $\exp(-\Delta\tau\hat{H})$, one may derive the relations $Z = \mathrm{Tr}\, \hat{G}\exp(-\beta\hat{H})$, $\hat{G} \equiv \prod_r \delta_{\hat{Q}_r,0}$, and $\hat{Q}_r \equiv \sum_i \nabla_i \hat{E}_{ri}$. Here, $\hat{Q}_r$ is the generator of the time-independent gauge transformation and $\hat{H}$ respects this symmetry as $[\hat{H}, \hat{Q}_r] = 0$. The Gauss's law $\hat{Q}_r = 0$ is to be imposed as a constraint for physical states. To respect $\hat{G}$ in Z, one needs to insert $\prod_r \delta_{Q_{(r,x_4)},0}$ at least only once at any x_4 in the path-integral due to $[\hat{H}, \hat{G}] = 0$. In other words, one may insert it at every x_4 as done in Eq. (4) due to the equalities, $\hat{G}^2 = \hat{G}$, $\hat{G}\exp(-\beta\hat{H}) = \hat{G}[\exp(-\beta\hat{H}/N)\hat{G}]^N$.

3. Atomic quantum simulators of U(1) lattice gauge theory

We now turn to discuss cold-atom quantum simulators of the LGTs[4-10] , specifically focusing on the system with Bose-Einstein condensates (BECs) in an OL.[4] The schematic illustration of the proposed setup by Zohar and Reznik[4] is shown in Fig. 1. As shown below, the vector potential and its conjugate electric field are represented by the local condensate phase operators $\hat{\theta}_{ri}$ and their conjugate number operators $\hat{\eta}_{ri}$. These observables live on the links of a three dimensional OL, and hence each link of the lattice is represented by a separate BEC. Local gauge invariance is generated by a certain two- and effective four-body interactions between the condensates through the overlap of the wave functions at the vertices of the lattices. For this purpose, six-component BECs, characterized by the field operators Ψ_a

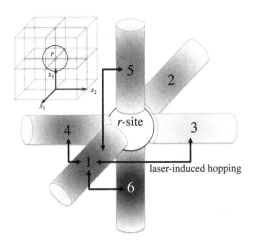

Fig. 1. Structure of the lattice. The localized wave functions of the six component BECs are concentrated on the links. At the vertex the wave functions of the neighboring links overlap and these are the only overlap integrals which are not negligible. The right panel shows a close-up picture of a single vertex. The laser-induced Raman coupling is only present between nearest neighbor orthogonal links; example for component 1 is only shown in the figure.

$(a = 1 \sim 6)$, are arranged to the links in the x_1-, x_2-, and x_3-directions in a manner shown in Fig. 1. Experimentally, this scheme can be implemented by using holographic masks techniques in order to generate the required OL^{17} . Note that the component suffices a directly connect to the direction of a link emanating from the site. This arrangement suppresses the usual hopping process between nearest-neighbor links, because location of the same component is well separated so that the overlapping of the wave functions in the different links is vanishing. Alternately, laser-induced Raman transitions are used to cause the controlled hopping with the transition (Rabi) frequency Ω between the nearest neighbor orthogonal links.

We start with a system of atomic condensates described by the Hamiltonian $H_{\mathrm{a}} = \int d\mathbf{r} \sum\limits_{a,b=1}^{6} H_{ab}(\mathbf{r})$, where

$$H_{ab} = \Psi_a^\dagger(\mathbf{r})\left[\delta_{ab}h_0^a(\mathbf{r}) + \Omega_{ab}\right]\Psi_b(\mathbf{r}) + \frac{G_{ab}}{2}\Psi_a^\dagger(\mathbf{r})\Psi_b^\dagger(\mathbf{r})\Psi_b(\mathbf{r})\Psi_a(\mathbf{r}), \quad (6)$$

where $a, b = 1 \sim 6$ represent the components of the wave function as well as the links emanating from each site. Here, δ_{ab} is Kronecker's delta, G_{ab} are the two-body coupling constants proportional to the s-wave scattering lengths and Ω_{ab} are Rabi frequencies. The single-particle Hamiltonian is given as $h_0^a(\mathbf{x}) = -\hbar^2\nabla^2/2m + V_a(\mathbf{r})$, where $V_a(\mathbf{r})$ is the OL of the component a. The laser-induced Rabi terms with Ω_{ab} couple the condensates to each other only in the L-shaped manner, as depicted in Fig. 1, and other terms are zero.

The second quantized wave functions of the condensates, taking into account only the lowest band excitations, are written as

$$\Psi_a(\mathbf{r}) = \sum_{r,a}{}' \hat{\psi}_{ra}\phi_{ra}(\mathbf{r}), \tag{7}$$

where $\hat{\psi}_{ra}$ are single-mode annihilation operator, annihilating one particle in the ground state of the link to a-direction emanating from the site r. Note that all the values r are not included in the sum of Eq. (7) because of the lattice's structure, distinguished by dash upon $\sum$. We assume that the local Wannier functions respect the symmetries $\phi_{ra}(\mathbf{r}) = \phi(\mathbf{r}-\mathbf{r}_a)$, $\phi_{ra}(\mathbf{r}) = \phi_{r,a+1}(\mathbf{r})$ for $a = 1, 3, 5$, and $\phi_{r1(2)}(\mathbf{r}) = \phi_{r3(4)}(R\mathbf{r}) = \phi_{r5(6)}(R'\mathbf{r})$, where R and R' are the appropriate rotation operators. The Wannier functions are chosen to be real without loss of generality.

Plugging the wave functions into the Hamiltonian (6), one can see that the only non-negligible contributions are the single-particle energy $\epsilon_0 \equiv \int d\mathbf{r}\phi(\mathbf{r} - \mathbf{r}_0)\left[-\hbar^2\nabla^2/2m + V_a(\mathbf{r})\right]\phi(\mathbf{r} - \mathbf{r}_0)$, the on-link integral $I_0 \equiv \int d\mathbf{r}\phi(\mathbf{r} - \mathbf{r}_0)^4$, the overlap integral between adjacent links in the same direction $I_s \equiv \int d^3\mathbf{r}\phi(\mathbf{r} - \mathbf{r}_0)^2\phi(\mathbf{r} - \mathbf{r}_s)^2$, the overlap integrals between adjacent links in the orthogonal direction $I_d \equiv \int d^3\mathbf{r}\phi(\mathbf{r} - \mathbf{r}_0)^2\phi(\mathbf{r} - \mathbf{r}_d)^2$ and $I_d' \equiv \int d^3\mathbf{r}\phi(\mathbf{r} - \mathbf{r}_0)\phi(\mathbf{r} - \mathbf{r}_d)$. Here, $\mathbf{r}_0$ is the position of an arbitrary link of the potential (due to the symmetries), and $\mathbf{r}_s$ and $\mathbf{r}_d$ are the positions of minima next to $\mathbf{r}_0$ in the same and orthogonal direction, respectively. The parameters of the system are now rewritten by including these integral contributions as $G_{aa}I_0 \equiv V_0$, $G_{ab}I_s = G_{ab}I_d \equiv g_{ab}$, and $\Omega_{ab}I_d' \equiv \Omega_{ab}'$.

We write the boson operator on the link as $\hat{\psi}_{ri} = \sqrt{\hat{\rho}_{ri}}$ $\exp[(-)^{x_1+x_2+x_3}i\hat{\theta}_{ri}]$, where we use the same letter θ_{ri} as $\theta_{x\mu}$ in Eq. (1) because the former is to be identified as the latter. We arrive at the following atomic Hamiltonian,[4]

$$\hat{H}_a = \sum_{r,a,b}{}' \left[g_{ab}\hat{\rho}_{ra}\hat{\rho}_{rb} + \frac{V_0}{2}\hat{\rho}_{ra}^2 + \Omega_{ab}'(\hat{\psi}_{ra}^\dagger\hat{\psi}_{rb} + \text{H.c.})\right]. \tag{8}$$

The g_{ab}-term describes the densty-density interaction, the $V_0(> 0)$-term is the on-site repulsion, and the Ω_{ab}'-term is the hopping term induced by external electromagnetic fields. We assume that the average $\langle\hat{\rho}_{ri}\rangle = \rho_0$ is homogeneous and large, $\rho_0 \gg 1$, and set $\hat{\rho}_{ri} = \rho_0 + (-)^{x_1+x_2+x_3}\hat{\eta}_{ri}$, where $\hat{\eta}_{ri}$ is the density fluctuation. Then, by choosing g_{ab} and Ω'_{ab} suitably[4-8] as $g_{ab} = 1/\gamma^2$ for any a and b, $\Omega'_{ab} \simeq 0$ for parallel link pairs, and $\Omega'_{ab} = \Omega'$ for perpendicular pairs, $\hat{H}_a$ is rewritten effectively as

$$\hat{H}_a = \frac{1}{2\gamma^2} \sum_r \Big(\sum_i \nabla_i \hat{\eta}_{ri}\Big)^2 + V_0 \sum_{r,i} \hat{\eta}_{ri}^2 + \hat{H}_{\mathrm{L}}(\{\hat{\theta}_{ri}\}) + O(\eta^3), \quad (9)$$

$$\hat{H}_{\mathrm{L}} = 2\Omega'\rho_0 \sum_{r,i<j} \Big[\cos(\hat{\theta}_{ri} - \hat{\theta}_{rj}) + \cos(\hat{\theta}_{ri} + \hat{\theta}_{r+i,j})$$

$$+ \cos(\hat{\theta}_{r+i,j} - \hat{\theta}_{r+j,i}) + \cos(\hat{\theta}_{rj} + \hat{\theta}_{r+j,i})\Big], \quad (10)$$

where we have neglected the overall constant. Eq. (9) is justified by (i) for $\eta_{xi}/\rho_0 \le 1$, we can neglect the additional $O(\eta^3)$ term originated from the hopping term, (ii) for $\eta_{xi}/\rho_0 \ge 1$, the path-integral has a damping factor of $O(e^{-\Delta\tau V_0\rho_0^2}) \ll 1$ for $V_0 = O(\rho_0^0)$. The term with γ^2 comes from the g_{ab}-term and represents the strength of the correlation of fluctuations $\hat{\eta}_{ra}$ around each site (partial conservation of atomic number). $\hat{H}_{\mathrm{L}}$ describes the phase correlation between the L-shaped nearest-neighbor links. We note that setting g_{ab} independent of a, b and controlling its magnitude γ^{-2} may be achieved by designing the OL suitably or by using optical Feshbach resonances[18-20] . Some theoretical ideas for the latter are also proposed.[21]

We use the coherent state $|\{\psi_{ri}\}\rangle$ and $\hat{1} = \prod_{r,i} \int d\rho_{ri}d\theta_{ri} \, |\{\psi_{ri}\}\rangle\langle\{\psi_{ri}\}|$ to obtain the path-integral for $Z_a = \mathrm{Tr} \exp(-\beta\hat{H}_a)$ as

$$Z_{\mathrm{a}} = \int \prod_{x,i} [d\eta_{xi}d\theta_{xi}] \exp\Big[\sum_{x,i}\Big(-i\eta_{xi}\nabla_4\theta_{xi} - \Delta\tau V_0\eta_{xi}^2\Big)$$

$$- \frac{\Delta\tau}{2\gamma^2} \sum_x \Big(\sum_i \nabla_i\eta_{xi}\Big)^2 - \Delta\tau \sum_{x_4} H_{\mathrm{L}}(\{\theta_{xi}\})\Big]. \quad (11)$$

The first term in the exponent in the rhs comes from $\sum_{x_4} \bar{\psi}_{xi}\nabla_4\psi_{xi} \simeq i\sum_{x_4} \eta_{xi}\nabla_4\theta_{xi}$ and shows that $-\hat{\eta}_{ri}$ is the conjugate momentum of $\hat{\theta}_{ri}$, whereby $\hat{E}_{ri} = -\hat{\eta}_{ri}$.

The Gaussian factor $\tilde{G} \equiv \prod_x \exp[(-\Delta\tau/2\gamma^2)Q_x^2]$ in Eq. (11) with $Q_x \equiv -\sum_i \nabla_i\eta_{xi}$ shows that the Gauss's law $Q_x = 0$ of Eq. (4) is achieved by $\tilde{G} \propto \prod_x \delta(Q_x)$ only at $\gamma \to 0$, and it is now shifted for $\gamma > 0$ to a Gaussian distribution with $Q_x^2 \lesssim \gamma^2/\Delta\tau$. Thus, γ is a parameter used to measure the violation of Gauss's law. Note that $\tilde{G}$ may be written as

$$\tilde{G} \simeq \int_0^{2\pi} \prod_x \frac{d\theta_{x4}}{2\pi} \exp\left(\frac{\gamma^2}{\Delta\tau}\cos\theta_{x4} - i\theta_{x4}\sum_i \nabla_i \eta_{xi}\right). \tag{12}$$

Substituting Eq. (12) to Eq. (11) and integrating over $\eta_{xi} \in (-\infty, \infty)$, one obtains a term $-(4\Delta\tau V_0)^{-1}(\nabla_4\theta_{xi} - \nabla_i\theta_{x4})^2$, which is a part of Gaussian Maxwell term. However, this result should be improved to respect the periodicity under $\theta_{xi} \to \theta_{xi} + 2\pi$, because θ_{xi} is the phase of the condensate. This Gaussian term is to be replaced, e.g., by a periodic Gaussian form or by the corresponding cosine form $\cos\theta_{xi4}$ as in Eq. (3) (which may be achieved by summing over the integer η_{xi}). After the summation over η_{xi}, Z_a may be expressed by the following general form;

$$Z_a = \int [dU] \exp(A_a), \quad A_a = A_I + A_P + A_L,$$

$$A_I = \sum_{x,\mu} c_{1\mu} \cos\theta_{x\mu}, \quad A_P = \sum_{x,\mu<\nu} c_{2\mu\nu} \cos\theta_{x\mu\nu},$$

$$A_L = \sum_{x,\mu<\nu} c_{3\mu\nu}\Big[\cos(\theta_{x\mu} - \theta_{x\nu}) + \cos(\theta_{x\mu} + \theta_{x+\mu,\nu})$$

$$+ \cos(\theta_{x+\mu,\nu} - \theta_{x+\nu,\mu}) + \cos(\theta_{x\nu} + \theta_{x+\nu,\mu})\Big]. \tag{13}$$

The anisotropic parameters in A_a are given as $c_{14} = \gamma^2/\Delta\tau$, $c_{1i} = 0$ and $c_{2i4} \simeq (2\Delta\tau V_0)^{-1}$; we discuss later how to implement nonzero c_{1i}. $\hat{H}_L$ with general values of Ω' directly gives rise to the A_L term with $c_{3i4} = 0$ and $c_{3ij} = 2\Omega'\rho_0\Delta\tau$, while $c_{2ij} = 0$.

We note that, for Ω' much smaller than γ^{-2} and/or V_0, one may treat $\hat{H}_L$ as a perturbation. In Refs.[4,5,8], the case $\gamma \simeq 0$ is considered to enforce the Gauss's law, and the second-order perturbation theory is invoked to obtain an anisotropic version of the KS Hamiltonian (5) as an effective Hamiltonian for the gauge-invariant subspace. This implies $c_{2ij} \simeq \gamma^2\rho_0^2\Omega'^2\Delta\tau$ and $c_{3\mu\nu} = 0$ in Eq. (13).

Figure 2 shows the phase diagrams of the model in Eq.(13) in the c_2-$c_{1,3}$ plane obtained by standard Monte Carlo simulations. Figure 2 (a) shows that the confinement phases of the pure gauge theory along $c_1 = c_3 = 0$ survive only up to the phase boundary $c_{1(3)} = c_{1(3)c}(c_2)$. Beyond this value of $c_{1(3)}$ the system enters into a new phase, "the Higgs phase" as explained below. The expectation that the cold atoms may simulate the pure gauge theory[4,5,8] is assured qualitatively and globally as long as both systems are in the same phase. This occurs for the atomic parameters satisfying

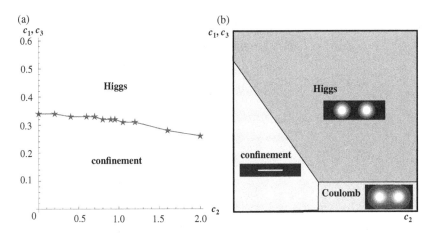

Fig. 2.   Global phase diagram of the U(1) gauge-Higgs model in the $c_2 - c_{1,3}$ plane. (a) Phase diagrams of the models (13), relevant to the atomic system, determined by $U = \langle A \rangle$ and $C = \langle A^2 \rangle - \langle A \rangle^2$ calculated by Monte Carlo simulations for a lattice size of 16^4. The vertical axis is $c_1 = c_3$ with $c_1 = c_{14}$ and $c_3 = c_{3ij}$ $(i, j = 1, 2, 3)$ and the horizontal axis is $c_{2i4} = c_2$ and $c_{2ij} = 0$. The confinement-Coulomb transition is missing in this model. (b) Typical phase diagram of the U(1) gauge-Higgs model in $c_2 - c_{1,3}$ plane. We also show contour plots of the deviation of typical atomic density $\Delta \rho_r \equiv (\sum_i \eta_{ri}^2 / 3)^{1/2}$ in the $x_1 - x_2$ plane at $x_3 = 0$ with external sources of atoms $\Delta \rho_{\text{ext}} = \pm \rho_1$ placed on the links emanating from $r = \pm r_0$. The white regions have $\Delta \rho_r$ greater than a certain value and the darker regions have lower $\Delta \rho_r$ The atomic density on the link $(r, r + i)$ is given by $\rho_{ri} = \rho_0 + \eta_{ri}$ (here we discard the factor $(-)^r$ in front of η_{ri} for simplicity), and the deviation η_{xi} is calculated by using the electric field $E_{ri}(= -\eta_{ri})$ with a pair of external sources $q = \pm 1$ at $r = r_\pm$. In the Higgs phase, $\Delta \rho_r$ decreases rapidly away from the sources. In the confinement phase, the deviation propagates from one source to the other along a one-dimensional string (electric flux).

$c_{1(3)} < c_{1(3)c}(c_2)$. This model has no Coulomb phase, which is consistent with the results of pure U(1) gauge theory that the confinement-Coulomb transition exists for 4D system[15] but not in the 3D system. It is known that the 3D version of the pure U(1) gauge theory Eq. (1) is always in the confinement phase (no Coulomb phase).[22]

4. Implementation of U(1) gauge-Higgs model

The A_{I} and A_{L} terms in Eq. (13) apparently break U(1) gauge invariance. However, the model Z_{a} of Eq. (13) with general set of parameters is equivalent to another LGT with exact U(1) gauge invariance, i.e., the U(1) gauge-Higgs model containing a Higgs field ϕ_x. The complex field ϕ_x is defined on site x and takes the form $\phi_x = \exp(i\varphi_x)$, whose radial excitation is frozen (so-called London limit). The partition function of the U(1) gauge-Higgs model Z_{GH} is defined by

$$Z_{GH} = \int [d\phi][dU] \exp A_{GH}(\{U_{x\mu}\}, \{\phi_x\}),$$

$$A_{GH} = A'_I + A_P + A'_L, \qquad \int [d\phi] \equiv \prod_x \int_0^{2\pi} \frac{d\varphi_x}{2\pi},$$

$$A'_I = \sum_{x,\mu} c_{1\mu} \cos(\varphi_x + \theta_{x\mu} - \varphi_{x+\mu}),$$

$$\begin{aligned}
A'_L = \sum_{x,\mu<\nu} c_{3\mu\nu} \Big[&\cos(\varphi_{x+\nu} + \theta_{x\mu} - \theta_{x\nu} - \varphi_{x+\mu}) \\
&+ \cos(\varphi_x + \theta_{x\mu} + \theta_{x+\mu,\nu} - \varphi_{x+\mu+\nu}) \\
&+ \cos(\varphi_{x+\mu} + \theta_{x+\mu,\nu} - \theta_{x+\nu,\mu} - \varphi_{x+\nu}) \\
&+ \cos(\varphi_x + \theta_{x\nu} + \theta_{x+\nu,\mu} - \varphi_{x+\nu+\mu}) \Big].
\end{aligned} \tag{14}$$

A_{GH} in Eq. (14) is gauge invariant under a simultaneous transformation of Eq. (2) and

$$\phi_x \equiv e^{i\varphi_x} \to \phi'_x = V_x \phi_x \qquad (\varphi_x \to \varphi'_x = \varphi_x + \Lambda_x). \tag{15}$$

In fact, Z_a is nothing but the gauge-fixed version of Z_{GH} with the so-called unitary gauge $\varphi_x = 0$. In short, the Higgs field ϕ_x represents a fictitious charged matter field to describe the violation of chargeless Gauss's law in ultra-cold atoms, where the general Gauss's law with a charged field is intact. This relation between a gauge-invariant Higgs model and its gauge-fixed version in the unitary gauge holds for a general action $\tilde{A}(\{U_{x\mu}\})$ as

$$\int [dU] e^{\tilde{A}(\{U_{x\mu}\})} = \int [dU][d\phi] e^{\tilde{A}(\{\bar{\phi}_{x+\mu} U_{x\mu} \phi_x\})}. \tag{16}$$

Eq. (16) is already known in high-energy physics where the standard U(1) gauge-Higgs model is the symmetric one, $c_{1\mu} = c_1$, $c_{2\mu\nu} = c_2$, $c_{3\mu\nu} = 0$, and used to discuss, e.g., the so-called complementarity relation between excitations in the confinement and Higgs phases.[23] However, its relevance to the quantum atomic simulator is quite important, because the relation $Z_a = Z_{GH}$ leads to a very interesting interpretation that the cold-atom systems proposed in Ref.[4] and the other related models[5-10] with a general set of values of parameters can be used as a simulator of a wider range of field theory, i.e., U(1) LGT including the Higgs couplings. For example, atomic simulations of the standard U(1) gauge-Higgs model above certainly open a new way to understand various phenomena including the inflation process of the early universe[13] and vortex dynamics of bosonized t-J model.[14]

As shown before, the magnitude of the Higgs coupling $c_{1\mu}$ of the atomic system such as the setup in Fig. 1 may be given as $c_{14} \neq 0$ but $c_{1i} = 0$. Concerning to c_{1i}, we note that nonvanishing c_{1i} terms may be incorporated into the cold-atom system by an idea discussed in Ref.[24] ; one may couple to $\hat{\psi}_{ri}$ the atomic field $\hat{a}_{ri}$ in another hyperfine state held in a different trapping potential via the interaction $\hat{H}_{a\psi} = \kappa \sum_{ri} \hat{a}_{ri}^{\dagger} \hat{\psi}_{ri} +$H.c. If $\hat{a}_{ri}$ condenses uniformly at sufficiently high temperatures, $\hat{a}_{ri}$ works as a BEC reservoir and $\hat{H}_{a\psi}$ supplies the c_{1i} term effectively with $c_{1i} = 2\kappa |\langle a_{ri} \rangle| / \sqrt{\rho_0} \Delta\tau$. A similar idea is also discussed in Ref.[25] to generate the c_{2ij} (spatial plaquette) term.

The global phase structure of the gauge-Higgs model Z_{GH} has generally similar structure shown in Fig. 2(b). There are generally three phases [see Fig. 2(b)] —Higgs, Coulomb, and confinement— in the order of increasing size of fluctuations of the gauge field $\theta_{x\mu}$. In the Higgs phase, both $\theta_{x\mu}$ and φ_x are stable. These three phases can be characterized by the potential energy $V(r)$ stored between two static charges with opposite signs and separated by a distance r, as $V(r) \propto 1/r$ (Coulomb), $\exp(-mr)/r$ (Higgs), r (confinement). One may distinguish each phase in the cold atom experiments by measuring atomic density [See Fig. 2(b)].

5. Summary and discussion

In summary, Eq. (14) is the target LGT of cold-atom systems that are basically those studied in Refs.[4,5] but with more general values of interaction parameters and a possible atomic reservoir.[24,25] From the discussion given in Refs.[4,5,24,25] and the relation (16), it may be rather universal that many cold-atom systems with multiplet "quantum spins" placed on OL links have their U(1) Higgs LGT counterparts. Such an equivalence between cold atoms and the U(1) gauge-Higgs model may be refered to as "quantum spin-gauge Higgs correspondence".

It is quite instructive to clarify the physical meaning of the Higgs phase of the gauge system realized in atomic quantum simulators. In the simulator using bosons,[4] the Higgs phase of the effective gauge system is nothing but the BEC state as the phase of the bosons (i.e., the gauge boson) is stabilized coherently. Therefore, the Higgs-confinement transition corresponds to the BEC transition. On the other hand, in Refs.[7,9] , the gauge field is expressed as $\hat{U}_{ri} \simeq (\hat{z}_{r+i}^{\sigma_r})^{\dagger} \hat{z}_r^{\sigma_r}$ ($\sigma_r = 1$ for even r and 2 for odd r) by using the Schwinger boson $\hat{z}_r^{\sigma}$, and the Higgs phase corresponds to the state in which the quantum state at each link $(r, r+i)$ is given by a coherent superposition of the particle-number states such as $|0\rangle_r|1\rangle_{r+i} + |1\rangle_r|0\rangle_{r+i}$. In the double-well potential, this state is realized naturally, after which the Higgs phase of the gauge system appears easily.

This way of introducing U(1) variables[7,9] reminds us of an approach starting with an antiferromagnet with $s = 1/2$ quantum spin at each site and obtaining the CP^1+U(1) LGT,[26] which has a Schwinger-boson (CP^1) variable at each site describing spins and an auxiliary but dynamical U(1) gauge variables on each link. Although the CP^1+U(1) model and the present U(1) Higgs model are different from each other, their global phase structures are significantly similar (See Fig. 1 of Ref.[26]).

Strictly speaking, Refs.[5-7] deal with the subspace $E_{ri} = 0, \pm1$ (the so-called U(1) gauge magnet or quantum link model) instead of $E_{ri} \in \mathbf{Z}$. There are some qualitative difference between these models and the KS model. The strength of confinement measured by the string tension α is weak as $\alpha \sim \exp(-1/g^2) \ll 1$ in the KS model for small $g^2 \ll 1$, where the gauge coupling is defined generally as $g^2 \propto c_2^{-1}$ in Eq.(5). On the other hand, the gauge magnet model corresponds to the case $g^2 \gg 1$ of the KS model in the sense that it allows only the $E_{ri} = 0, \pm1$ states, and the confinement is strong as $\alpha \sim g^2 \gg 1$, which may be calculated by the strong-coupling expansion. For a large V_0, which corresponds to $c_2 \ll 1$ ($c_{2i4} \propto 1/V_0$), the two models may have similar behaviors, because Eq.(5) with $c_2 \ll 1$ restricts E_{ri} to $0, \pm1$ effectively. Comparison of these two models in a quantitative and general setting is an interesting problem.

References

1. M. Lewenstein, A. Sanpera, and V. Ahufinger, *Ultracold Atoms in Optical Lattices: Simulating Quantum Many-body Systems* (Oxford University Press, 2012).
2. U. -J. Wiese, arXiv:1305.1602 (2013).
3. N. Goldman, G. Juzeliunas, P. Ohberg, I. B. Spielman, arXiv:1308.6533.
4. E. Zohar and B. Reznik, Phys. Rev. Lett. **107**, 275301 (2011).
5. E. Zohar, J. I. Cirac, and B. Reznich, Phys. Rev. Lett. **109**, 125302 (2012).
6. L. Tagliacozzo, A. Celi, A. Zamora, and M. Lewenstein, Ann. Phys. **330**, 160 (2013).
7. D. Banerjee, M. Dalmonte, M. Müller, E. Rico, P. Stebler, U.-J. Wiese, and P. Zoller, Phys. Rev. Lett. **109**, 175302 (2012).
8. E. Zohar, J. I. Cirac, and B. Reznich, Phys. Rev. Lett. **110**, 055302 (2013).
9. E. Zohar, J. I. Cirac, and B. Reznich, Phys. Rev. Lett. **110**, 125304 (2013).
10. D. Banerjee, M.Bögli, M. Dalmonte, E. Rico, P. Stebler, U.-J. Wiese, and P. Zoller, Phys. Rev. Lett. **110**, 125303 (2013).
11. K. Wilson, Phys. Rev. D **10**, 2445 (1974); J. B. Kogut, Rev. Mod. Phys. **51**,

659 (1979).

12. K. Kasamatsu, I. Ichinose, and T. Matsui, Phys. Rev. Lett. **111**, 115303 (2013).

13. A. H. Guth, Phys. Rev. D **23**, 347 (1981); E. Kolb and M. Turner, "The Early Universe", Westview Press, (1994); A. D. Linde, Lect. Notes Phys. **738**, 1 (2008).

14. See, e.g., K. Aoki, K. Sakakibara, I. Ichinose and T. Matsui, Phys. Rev. B **80**, 144510 (2009).

15. See, e.g., E. Sa'nchez-Velasco, Phys. Rev. E **54**, 5819 (1996), and references cited therein.

16. J. Kogut and L. Susskind, Phys. Rev. D **11**, 395 (1975).

17. K. Dholakia and T. Cizmar, Nature Photon **5**, 335 (2011).

18. P. O. Fedichev, Y. Kagan, G. V. Shlyapnikov, and J. T. M. Walraven, Phys. Rev. Lett. **77**, 2913 (1996).

19. J. L. Bohn and P. S. Julienne, Phys. Rev. A **56**, 1486 (1997).

20. F. K. Fatemi, K. M. Jones, and P. D. Lett, Phys. Rev. Lett. **85**, 4462 (2000).

21. P. Zhang, P. Naidon, and M. Ueda, Phys. Rev. Lett. **103**, 133202 (2009).

22. A. M. Polyakov, Phys. Lett. B **59**, 82 (1975).

23. E. Fradkin and S. H. Shenker, Phys. Rev. D **19**, 3682 (1979).

24. A. Recati, P. O. Fedichev, W. Zwerger, J. von Delft, and P. Zoller, Phys. Rev. Lett. **94**, 040404 (2005).

25. H. P. Büchler, M. Hermele, S. D. Huber, M. P. A. Fisher, and P. Zoller, Phys. Rev. Lett. **95**, 040402 (2005).

26. K. Sawamura, T. Hiramatsu, K. Ozaki, I. Ichinose, and T. Matsui, Phys. Rev. B **77**, 224404 (2008).

RECURSIVE CONSTRUCTION OF NOISELESS SUBSYSTEM FOR QUDITS

UTKAN GÜNGÖRDÜ[1], CHI-KWONG LI[2], MIKIO NAKAHARA[1,3], YIU-TUNG
POON[4] AND NUNG-SING SZE[5]

[1] *Research Center for Quantum Computing, Interdisciplinary Graduate School of
Science and Engineering, Kinki University, 3-4-1 Kowakae, Higashi-Osaka, Osaka
577-8502, Japan*

[2] *Department of Mathematics, College of William & Mary, Williamsburg, VA
23187-8795, USA. (Year 2011: Department of Mathematics, Hong Kong University of
Science & Technology, Hong Kong.)*

[3] *Department of Physics, Kinki University, 3-4-1 Kowakae, Higashi-Osaka, Osaka
577-8502, Japan*

[4] *Department of Mathematics, Iowa State University, Ames, IA 50011, USA*

[5] *Department of Applied Mathematics, The Hong Kong Polytechnic University, Hung
Hom, Hong Kong*

When the environmental noise acting on the system has certain symmetries,
a subsystem of the total system can avoid errors. Encoding information into
such a subsystem is advantageous since it does not require any error syndrome
measurements, which may introduce further errors to the system. However,
utilizing such a subsystem for large systems gets impractical with the increasing
number of qudits. A recursive scheme offers a solution to this problem. Here,
we review the recursive construct introduced in,[1] which can asymptotically
protect $1/d$ of the qudits in system against collective errors.

Keywords: Quantum information processing; Quantum error correction; Decoherence; Representation theory.

1. Introduction

Quantum computation and quantum information processing rely on storing
and processing the information using a quantum system. One of the most
important obstacles against realization of a practical computer is decoherence, a process that is caused by the coupling of the quantum system to
an external environment. A promising way of fighting against the effects
of decoherence is quantum error correction codes (QECC).[2] One can introduce redundancy in the system initially, and make use of it to identify
the errors by performing measurements and checking the error syndrome.

However, a measurement may introduce further errors to the system. When the system-environment interaction causes errors only in a part of the total Hilbert space, one can encode the information into this subspace and avoid errors by construction. Decoherence free subspace (DFS) and noiseless subsystem (NS) are two well-known examples of such "error-avoiding" schemes.[2-13]

When all qubits suffer from the same error operator, that is to say, when the environmental noise possesses permutation symmetry, the system is said to be under the influence of a collective noise.[14-16] This can happen when the size of the quantum system is much smaller than the wavelength of the environmental disturbance or when photonic qubits are transmitted through an optical fiber that has a fixed imperfection. Another instance where the collective noise is relevant is when Alice sends qubits to Bob without knowing the basis vectors Bob is using for measurements. Because the mismatching of the vectors would be common for all qubits, it can be regarded as a collective noise.

In,[17] it was shown that when the external disturbance is limited to the generators of $SU(2)$ such that the error operators belong to the set $\{\sigma_x^{\otimes n}, \sigma_y^{\otimes n}, \sigma_z^{\otimes n}\}$, one can iteratively implement a QECC which can protect $n-1$ logical qubits out of n physical qubits when n is odd, and $n-2$ logical qubits when n is even. In this scheme, the encoding rate, which is defined as the ratio of the number of protected qubits to the number of total qubits, is asymptotically 1 for even and odd cases.

For the more general case where the error operators are not restricted to a finite set, that is $\mathcal{E} = W^{\otimes n}$ where W is an arbitrary element of $SU(2)$, an explicit recursive method has been described in.[18] In this scheme $n = 2k+1$ qubits is required to protect k logical qubits, thus the asymptotic encoding rate is $1/2$.

A more general setting involves qudits, which are d-dimensional generalization of qubits (and should not be confused with d-dimensional representations of qubits, such as spin-1 systems). They transform under the action of the elements of $SU(d)$ [a]. In,[19] the largest subspace that is immune to collective noise $\mathcal{E} = W^{\otimes n}$ (where $W \in SU(d)$) has been identified for qubits ($d = 2$) and qutrits ($d = 3$). It was shown that the asymptotic encoding rate of such subspaces approach to 1 when $n \to \infty$. However, the problem for $d > 3$ is an open and highly non-trivial one even though the

---

[a]Strictly speaking, qudits transform under the action of the unitary group $U(d)$. However, in quantum mechanics, a global phase is not of practical importance, thus we prefer to work with the special unitary group $SU(d) \cong U(d)/U(1)$.

decomposition rules of the collective noise operator are well-established.

This work, which is based on,[1] analyzes[18] from a representation theory point of view, explains how the recursive scheme works and generalizes the results qudits. The resulting asymptotic encoding rate of this scheme turns out to be $1/d$, and the encoding/decoding can be realized by operating only on $d + 1$ neighboring qudits at a time. A natural question for this results is "why do we do this analysis even though we know that there exists an encoding that gives asymptotic encoding rate of 1?". With growing n, finding the encoding/decoding circuit $U_E/U_E^\dagger$ becomes impractical. For 100 or even 10 qudits, it is totally impossible to find the required quantum circuit. We believe a recursive construction is the only practical way to realize DFS/NS for large n.

2. The recursive scheme for qubits

In this section, we revisit the recursive scheme for qubits described in,[18] which has an encoding rate of $1/d$. We do this by employing ideas from the representation theory, which gives a broad and general view.

The physical system we consider here is an array of three qubits. Each site is assumed to be affected by the same error operator, W, thus the collective error is $\mathcal{E} = W \otimes W \otimes W$. Such a noise is symmetric under all permutations involving three particles, namely, it belongs to the $S_3 \supset S_2$ symmetry group chain (here, S_n denotes the symmetric group of n particles).

In the language of representation theory, irreps of groups are traditionally denoted using Young tableaux. The fundamental irrep of SU(2), in particular, is labeled as $\boxed{1}$. Tensor product of 3 fundamental irreps are reduced in terms of higher irreps in accordance with[20]

$$\boxed{1} \otimes \boxed{1} \otimes \boxed{1} = \boxed{\begin{array}{cc} 1 & 2 \\ \hline 3 \end{array}} \oplus \boxed{\begin{array}{cc} 1 & 3 \\ \hline 2 \end{array}} \oplus \boxed{\begin{array}{ccc} 1 & 2 & 3 \end{array}}. \tag{1}$$

The irreps appearing on the RHS have the dimensions of 2, 2 and 4 respectively thus the above equation is sometimes written as $\mathbf{2 \times 2 \times 2 = 2 + 2 + 4}$. This means the collective error operator $\mathcal{E}$ will be block-diagonalized with the block structure of 2, 2, 4 when reduced. Notice that the irreps $\boxed{1}$, $\begin{array}{cc} 1 & 2 \\ \hline 3 \end{array}$ and $\begin{array}{cc} 1 & 3 \\ \hline 2 \end{array}$ all have the dimension of 2 —for the case of SU(d), one can trim the leftmost rectangular block of height d when calculating the dimension and all such irreps are completely equivalent. The two copies of the fundamental irrep on the RHS is the redundancy that eventually gives rise to a noiseless subsystem.

When $\mathcal{E}$ is block-diagonalized, each block represents the same SU(2) rotation as W, only in a higher dimensional space. The 4×4 block for instance represents the spin-3/2 [b] version of the original spin-1/2 rotation $W = e^{i(r_x \sigma_x + r_y \sigma_y + r_z \sigma_z)}$, and is expressed as

$$W^{(3/2)} = e^{i[r_x J_x^{(3/2)} + r_y J_y^{(3/2)} + r_z J_z^{(3/2)}]}, \tag{2}$$

in terms of the 4-dimensional generators $J_i^{(3/2)}$ (see for example[21,22] for the higher dimensional representations of the $\mathfrak{su}(2)$ algebra)

$$J_x^{(3/2)} = \begin{pmatrix} 0 & \sqrt{3} & 0 & 0 \\ \sqrt{3} & 0 & 2 & 0 \\ 0 & 2 & 0 & \sqrt{3} \\ 0 & 0 & \sqrt{3} & 0 \end{pmatrix}, J_y^{(3/2)} = i \begin{pmatrix} 0 & -\sqrt{3} & 0 & 0 \\ \sqrt{3} & 0 & -2 & 0 \\ 0 & 2 & 0 & -\sqrt{3} \\ 0 & 0 & \sqrt{3} & 0 \end{pmatrix},$$

$$J_z^{(3/2)} = \begin{pmatrix} 3 & 0 & 0 & 0 \\ 0 & 1 & 0 & 0 \\ 0 & 0 & -1 & 0 \\ 0 & 0 & 0 & -3 \end{pmatrix}. \tag{3}$$

Let us denote the elements of the fundamental irrep as u and d (we will also use the braket notation $|u\rangle$, $|d\rangle$ interchangeably), which allude to the spin-up/down states of the electron. The vectors that belong to the irreps appearing on the RHS of Eq. (1) can be written as

$$\boxed{\begin{array}{|c|c|} \hline 1 & 2 \\ \hline 3 \\ \hline \end{array}} \begin{cases} \dfrac{1}{\sqrt{6}}(-[ud+du]u + 2[uu]d) \\ \dfrac{1}{\sqrt{6}}(2[dd]u - [ud+du]d) \end{cases}$$

$$\boxed{\begin{array}{|c|c|} \hline 1 & 3 \\ \hline 2 \\ \hline \end{array}} \begin{cases} \dfrac{1}{\sqrt{2}}(ud-du)u \\ \dfrac{1}{\sqrt{2}}(ud-du)d \end{cases} \tag{4}$$

$$\boxed{\begin{array}{|c|c|c|} \hline 1 & 2 & 3 \\ \hline \end{array}} \begin{cases} uuu \\ \dfrac{1}{\sqrt{3}}(uud+udu+duu) \\ \dfrac{1}{\sqrt{3}}(ddu+dud+udd) \\ ddd \end{cases}$$

---

[b]In the language of angular momentum, this would be called $l = 3/2$ representation.

The coefficients appearing in these vectors are called Clebsch-Gordan coefficients, and the vectors are sometimes called Young-Yamanouchi vectors.[23] The unitary matrix U_E that block-diagonalizes $\mathcal{E}$ as $\mathcal{E}' = U_E^\dagger \mathcal{E} U_E = W \oplus W \oplus W^{(3/2)}$ is constructed by using these basis vectors as columns and grouping them in a proper fashion such as

$$U_E = \begin{pmatrix} \vdots & \vdots & & \vdots \\ \boxed{\begin{array}{cc}1&2\\\hline 3\end{array}} & \boxed{\begin{array}{cc}1&3\\\hline 2\end{array}} & \boxed{\begin{array}{ccc}1&2&3\end{array}} \\ \vdots & \vdots & & \vdots \end{pmatrix}, \tag{5}$$

where vertical dots indicate that vectors of the irrep are placed as column vectors.

In Fig. 1 the full operation of sending the state $|u\rangle\,|\psi\rangle\,|a\rangle$ is depicted. The information is encoded into $|\psi\rangle$, and $|a\rangle$ is an arbitrary ancillary qubit.

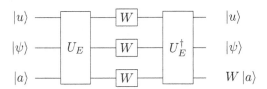

Fig. 1. Three qubit QECC from[18] (re-ordered version). Encoding operation U_E is given in Fig. 2.

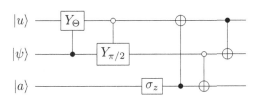

Fig. 2. The encoding gate U_E from[18] (re-ordered), in terms of single-qubit and two qubit controlled-U gates. Above $Y_\theta = \exp(i\sigma_y\theta)$ and $\sin\Theta = \sqrt{2/3}$. Such an expansion in terms of "elementary" gates is not unique.

During the transmission, only the ancillary qubit $|a\rangle$ is affected. This fact, and the particular choice of the input state can be explained like this. An appropriate density matrix ρ for the quantum channel with collective noise is determined by the form of the reduced error operator

$\mathcal{E}' = U_E^\dagger \mathcal{E} U_E = (\mathbb{1}_2 \otimes W) \oplus W^{(3/2)}$. When ρ is chosen as $(|\psi\rangle\langle\psi| \otimes |a\rangle\langle a|) \oplus 0_4$, clearly, only the ancillary qubit is distorted by W while remainder of the system is left unharmed [c].

An important observation here is that, the action of $\mathcal{E}'$ on the subspace spanned by $|u\rangle |\psi\rangle |a\rangle$ is equivalent to $\mathbb{1}_2^{\otimes 2} \otimes W$. This fact can be used to construct a noiseless subsystem for $2k + 1$ qubits in a recursive manner as follows. Suppose that we want to construct a 5-qubit noiseless subsystem. Due to this equivalence, the circuit given in Figs. 3 and 4 are identical. Moreover, one can take this one step further and reduce the circuit given in Fig. 4 into the one given in Fig. 5, where it is clear that only the bottom qubit is affected during the transmission. This inductively illustrates the construction of a noiseless subsystem for $2k + 1$ qudits, that can protect k logical qubits from environmental noise.

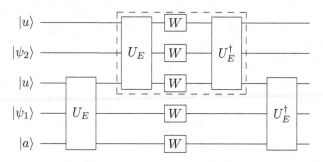

Fig. 3. A re-ordered version of the recursive 5-qubit circuit diagram from.[18] Due to re-ordering, the gates act on neighboring 3 qubits only.

3. General case: recursive construction of noiseless subsystem for qudits

The representation theory-based approach used in the previous section is general and can be extended to qudits in a straightforward manner. In this section, we describe an analogous elementary circuit similar to the one given in Fig. 1 and a recursive scheme (see Figs. 3, 4, 5) for d-level systems.

To proceed, we first determine the number of qudits m required for the elementary circuit, which can protect a single qudit from the collective noise $\mathcal{E} = W^{\otimes m}$. Here, W is an arbitrary, unknown error operator acting on a

---

[c]Here 0_m is an $m \times m$ zero matrix and $\mathbb{1}_m$ is an $m \times m$ identity matrix.

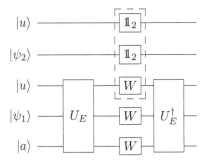

Fig. 4. This reduced circuit is equivalent to the one shown in Fig. 3 due to the equivalence given in the text.

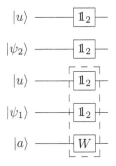

Fig. 5. The final version of the circuit given in Fig. 3.

single site, and is an element of $SU(d)$. In order to construct a noiseless subsystem, we must have sufficient redundancy in the system, determined by the multiplicity of irreps. The multiplicity of an irrep is equivalent to the dimension of the corresponding irrep of the symmetric group, which is given by the Frobenious formula.[23] As it turns out, the fundamental irrep $\boxed{1}$ appears exactly d times for $d+1$ qudits, which can be seen by using the Frobenious formula:

$$(d+1)! \frac{\prod_{1 \leq i < j \leq d} \nu_i - \nu_j + j - i}{\prod_{i=1}^{d} (\nu_i + d - i)!} = d. \tag{6}$$

Here, ν_i denotes the length of the ith row (top-to-bottom ordering) in the corresponding Young diagram. We utilize this multiplicity as follows. The encoding matrix U_E, constructed by placing the Young-Yamanouchi vectors[d] of the corresponding irreps as columns below

$$U_E = \begin{pmatrix} \vdots & \vdots & \vdots & \vdots & & \vdots \\ \boxed{1} & \boxed{1} & \cdots & \boxed{1} & \text{other irreps} \\ \vdots & \vdots & \vdots & \vdots & & \vdots \end{pmatrix}, \tag{7}$$

will reduce (block-diagonalize) the collective error operator $\mathcal{E}$ as

$$\mathcal{E}' = U_E^\dagger \mathcal{E} U_E = (\mathbb{1}_d \otimes W) \oplus \mathcal{O}. \tag{8}$$

Here, each $\boxed{1}$ denotes any irrep that is equivalent to the d-dimensional fundamental irrep. There are d such irreps in total. $\mathcal{O}$ represents the direct sum of other representations of W, which are not relevant for our purposes. The ordering of $\boxed{1}$ in U_E is not important, and the part involving other irreps can be chosen arbitrarily as long as the vectors are chosen to be orthogonal to the ones in the fundamental irreps which is enforced by the unitarity of U_E. These can be seen as a freedom during the construction of the encoding/decoding circuit in a practical setting.

This noiseless subsystem can protect the one-qudit state $|\psi\rangle$ when encoded into the input state as

$$|\Psi\rangle = |u\rangle^{\otimes d-1} |\psi\rangle |a\rangle, \tag{9}$$

where $|a\rangle$ is an arbitrary ancillary state and $|u\rangle$ is the highest-weight state in the fundamental irrep of $SU(d)$, the d-dimensional vector $(1, 0, \ldots, 0)^T$. When the encoding/decoding is seen as a unitary transformation, it becomes clear that $|\Psi\rangle$ belongs to the direct-sum space of the fundamental irreps, as it has non-zero entries only in the first d^2 rows.

The density matrix of the system $\rho = |\Psi\rangle \langle\Psi|$ is $(|\psi\rangle \langle\psi| \otimes |a\rangle \langle a|) \oplus 0_q$ where $q = d^{d+1} - d^2$. The action of the collective noise $\mathcal{E}'$ given in Eq. (8) on this state only affects the ancillary qudit, by distorting it into $W |a\rangle$. All the remaining qudits are unaffected during the transmission through the noisy channel. We observe the the action of $\mathcal{E}'$ on this subspace is equivalent to $\mathbb{1}_d^{\otimes d} W$, or that the equivalence

$$(U_E^\dagger W^{\otimes d+1} U_E) |u\rangle^{\otimes d-1} |\psi\rangle |a\rangle = |u\rangle^{\otimes d-1} |\psi\rangle (W |a\rangle) \tag{10}$$

holds. Following the arguments on Figs. 3, 4 and 5, we see that this equivalence enables a recursive construction of a $kd + 1$-qudit QECC, which is capable of protecting k qudits.

---

[d] The Young-Yamanouchi vectors are constructed from $SU(d)$ Clebsch-Gordan coefficients. Details on their computation can be found in.[23,24]

A naive way of constructing noiseless subsystem for k qudits would be to vertically clone the elementary circuit such as the one given in Fig. 1. Since the elementary circuit protects a single qudit using $d + 1$ qudits, the asymptotic encoding rate would be $1/(d + 1)$. However, with the recursive scheme, given that the number of correctable qudits using $n = kd + 1$ for the channel is k, we find the asymptotic behavior of the encoding rate to be $k/n \to 1/d$ as $n \to \infty$ for a fixed d.

4. Concluding remarks

The noiseless subsystem is a method of using the inherent permutation symmetry of a quantum channel to protect a subsystem against errors. In this work, we have used various tools from representation theory for a better understanding and further generalization of the recursive construction of a noiseless subsystem for qubits[18] and extended our results to qudits. Our approach here is based on a $d + 1$-qudit encoding circuit whose implementation is realized by the basis vectors belonging to the fundamental irrep $\boxed{1}$. In what follows, we generalized this construction to $n = kd + 1$. Encoding/decoding operations can be realized by using $U_E/U_E^\dagger$ successively, operating on $d + 1$ *neighboring* qudits at a time, a fact which can be of practical importance.

However, we note that our construction does not give the maximum number of correctable qudits for a channel. When the irrep with maximal degeneracy is used instead of the fundamental representation, the ratio of protected qudits and total number of qudits is $k/n \to 1$ as $n \to \infty$.[19] Even though the DFS/NS with the maximal dimension is identified, we do not yet know how to implement the encoding circuit efficiently yet. Our study here gives a foolproof implementation of the encoding circuit although the efficiency is $1/d$ for qudits. It is certainly desirable to find a recursion relation for maximum dimensional DFS/NS, which we leave as a future work.

Finally, we remark that our scheme is applicable to the more general case of non-unitary error channels as well. The essential ingredient for the construction we described is the permutation symmetry of the noisy quantum channel whose action is given by $\mathcal{E} = W^{\otimes n}$. The Kraus operator W may well belong to a Lie group other than $SU(d)$, such as the special linear group $SL(d, \mathbb{C})$.

Acknowledgments

We would like to thank Paolo Zanardi, Daniel Lidar and Lorenza Viola for bringing some of the references to our attention. UG and MN are grateful to JSPS (Japan Society for the Promotion of Science) for partial support from Grant-in-Aid for Scientific Research (Grant Nos. 23540470 and 24320008). UG acknowledges the financial support of the MEXT (Ministry of Education, Culture, Sports, Science and Technology) Scholarship for foreign students. C.-K.L. was supported by a USA NSF grant, a HK RGC grant, and the 2011 Shanxi 100 Talent Program. He is an honorary professor of University of Hong Kong, Taiyuan University of Technology, and Shanghai University. Y.-T.P. was supported by a USA NSF grant and a HK RGC grant. N.-S.S. was supported by a HK RGC grant PolyU 502512.

References

1. U. Güngördü, C.-K. Li, M. Nakahara, Y.-T. Poon and N.-S. Sze, *arXiv preprint arXiv:1310.4401, submitted to journal* (2013).
2. P. Zanardi and M. Rasetti, *Physical Review Letters* **79**, 3306(October 1997).
3. P. Zanardi and M. Rasetti, *Modern Physics Letters B* **11**, p. 1085 (1997).
4. P. Zanardi, *Physical Review A* **57**, 3276(May 1998).
5. D. A. Lidar, I. L. Chuang and K. B. Whaley, *Physical Review Letters* **81**, 2594(September 1998).
6. J. Kempe, D. Bacon, D. A. Lidar and K. B. Whaley, *Physical Review A* **63**, p. 042307(March 2001).
7. D. A. Lidar, D. Bacon, J. Kempe and K. B. Whaley, *Physical Review A* **61**, p. 052307(April 2000).
8. D. A. Lidar, D. Bacon, J. Kempe and K. B. Whaley, *Physical Review A* **63**, p. 022306(January 2001).
9. M.-D. Choi and D. W. Kribs, *Phys. Rev. Lett.* **96**, p. 050501(February 2006).
10. E. Knill, R. Laflamme and L. Viola, *Physical Review Letters* **84**, 2525(March 2000).
11. E. M. Fortunato, L. Viola, M. A. Pravia, E. Knill, R. Laflamme, T. F. Havel and D. G. Cory, *Phys. Rev. A* **67**, p. 062303(June 2003).
12. L. Viola, E. M. Fortunato, M. a. Pravia, E. Knill, R. Laflamme and D. G. Cory, *Science* **293**, 2059(September 2001).
13. D. A. Lidar and T. A. Brun, *Quantum Error Correction* (Cambridge University Press, 2013).
14. L. Viola, E. M. Fortunato, M. a. Pravia, E. Knill, R. Laflamme and D. G. Cory, *Science (New York, N.Y.)* **293**, 2059(September 2001).
15. P. G. Kwiat, *Science* **290**, 498(October 2000).
16. D. Kielpinski, *Science* **291**, 1013(January 2001).
17. C.-K. Li, M. Nakahara, Y.-T. Poon, N.-S. Sze and H. Tomita, *Physics Letters A* **375**, 3255(August 2011).

18. C.-K. Li, M. Nakahara, Y.-T. Poon, N.-S. Sze and H. Tomita, *Physical Review A* **84**, p. 044301(October 2011).
19. C.-K. Li, M. Nakahara, Y.-T. Poon and N.-S. Sze, *arXiv preprint arXiv:1306.0981* (2013).
20. H. Georgi, *Lie Algebras In Particle Physics: from Isospin To Unified Theories*, 2nd edn. (Westview Press, 1995).
21. W. Pfeifer, *The Lie Algebras su(N): An Introduction* (Birkhäuser, 2003).
22. J. J. Sakurai, *Modern Quantum Mechanics*, 2nd edn. (Addison Wesley, 2010).
23. J.-Q. Chen, J. Ping and F. Wang, *Group Representation Theory For Physicists*, 2nd edn. (World Scientific, 2002).
24. A. Alex, M. Kalus, A. Huckleberry and J. von Delft, *Journal of Mathematical Physics* **52**, p. 023507 (2011).

COMPOSITE QUANTUM GATES FOR PRECISE QUANTUM CONTROL

MASAMITSU BANDO[*,1], TSUBASA ICHIKAWA[2], YASUSHI KONDO[1,3] and
MIKIO NAKAHARA[1,3]

[1] *Interdisciplinary Graduate School of Science and Engineering, Kinki University,
Higashiosaka, Osaka 577-8502, Japan*
[2] *Department of Physics, Gakushuin University, 1-5-1 Mejiro, Toshima-ku, Tokyo
171-8588, Japan*
[3] *Department of Physics, Kinki University, Higashiosaka, Osaka 577-8502, Japan*
[*] *E-mail: bando@alice.math.kindai.ac.jp*

In NMR quantum computation and NMR experiments, there are many composite quantum gates to suppress one of two dominant errors. In this paper, we review some concatenated composite pulses (CCCPs) that is robust against two dominant errors simultaneously.

In NMR quantum computation,[1] a single-qubit gate without errors takes the form

$$R(\theta, \phi) = \exp[-\mathrm{i}\theta \boldsymbol{n}(\phi) \cdot \boldsymbol{\sigma}/2], \tag{1}$$

where θ, ϕ and $\boldsymbol{n}(\phi) = (\cos\phi, \sin\phi, 0)$ are the rotation angle, the azimuthal angle and the rotation axis which specifies, respectively. And $\boldsymbol{\sigma} = (\sigma_x, \sigma_y, \sigma_z)$. In this paper, let us call $R(\theta, \phi)$ of the form (1) an elementary gate. There are two dominant systematic errors in NMR; a pulse length error (PLE), and an off-resonance error (ORE).[1,2] Under PLE and ORE, a single-qubit gate is perturbed as

$$R'(\theta, \phi) = \exp[-\mathrm{i}(1 + \varepsilon)\theta(\boldsymbol{n}(\phi) + f\hat{\boldsymbol{z}}) \cdot \boldsymbol{\sigma}/2]$$
$$= R(\theta, \phi) - \mathrm{i}\varepsilon\theta\boldsymbol{n}(\phi) \cdot \boldsymbol{\sigma}/2 - \mathrm{i}f\sin(\theta/2)\sigma_z + \mathcal{O}(\varepsilon^2, f^2, \varepsilon f) \tag{2}$$

where $\hat{\boldsymbol{z}} = (0, 0, 1)$, ε and f are unknown but fixed small constants that represent the strength of PLE and ORE, respectively.

A composite quantum gate is a gate sequence that is robust against PLE or ORE.[1,3,4] Let $U(\theta, \phi)$ denote a gate sequence of N elementary gates in

238

error-free case as

$$U(\theta, \phi) = R(\theta_N, \phi_N)R(\theta_{N-1}, \phi_{N-1}) \cdots R(\theta_1, \phi_1), \qquad (3)$$

and $U'(\theta, \phi)$ denote a gate sequence of N elementary gates under PLE and ORE as

$$U'(\theta, \phi) = R'(\theta_N, \phi_N)R'(\theta_{N-1}, \phi_{N-1}) \cdots R'(\theta_1, \phi_1), \qquad (4)$$

where the error strengths ε and f in all the elementary gates are the same and the set of elementary gates $\{R(\theta_i, \phi_i)\}$ is designed so as to satisfy $U(\theta, \phi) = R(\theta, \phi)$. The $U(\theta, \phi)$ is called a composite quantum gate if $U'(\theta, \phi) = R(\theta, \phi) + \mathcal{O}(\varepsilon^2, f^2, \varepsilon f)$ is satisfied. Figure 1 shows a image of structure of composite quantum gate. CORPSE[5] and BB1[6] are well known composite quantum gates that are robust against ORE and PLE, respectively. The CORPSE consists of three elementary gates with parameters

$$\begin{aligned}
\theta_1 &= \theta_3 + 2\pi = 2\pi + \theta/2 - k, \quad \theta_2 = 2\pi - 2k, \\
\phi_1 &= \phi_2 - \pi = \phi_3 = \phi, \quad k = \arcsin[\sin(\theta/2)/2],
\end{aligned} \qquad (5)$$

while the BB1 consists of three elementary gates with parameters

$$\begin{aligned}
\theta_1 &= \theta_3 = \pi, \quad \theta_2 = 2\pi, \quad \theta_4 = \theta, \\
\phi_1 &= \phi_3 = \phi + \arccos[-\theta/(4\pi)], \quad \phi_2 = 3\phi_1 - 2\phi, \quad \phi_4 = \phi.
\end{aligned} \qquad (6)$$

The gate fidelity of $R'(\theta, \phi)$ is defined as

$$F = |\mathrm{tr}(R^\dagger(\theta, \phi)R'(\theta, \phi))|/2. \qquad (7)$$

Composite quantum gates have no first order error terms of PLE and ORE and therefore they have high gate fidelity more than an elementary gate if $|\varepsilon| \ll 1$ and $|f| \ll 1$ are satisfied. Incidentally, all composite quantum gates that are robust against PLE are geometric quantum gates.[7,8] geometric quantum gates[9,10] based on the holonomy[11-14] are expected to be robust against random noises.

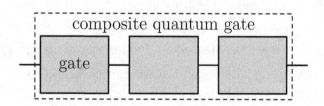

Fig. 1.　Structure of composite quantum gate.

A concatenated composite pulse (CCCP) is a nested composite quantum gate that is robust against PLE and ORE simultaneously.[15,16] To design a CCCP, we must choose two composite quantum gates; an outer composite quantum gate and inner composite quantum gate. Figure 2 explains this naming convention. For example, if we choose BB1 as outer composite quantum gate and CORPSE as inner composite quantum gate, we can design a CCCP that we call CORPSE in BB1 CCCP (CinBB).

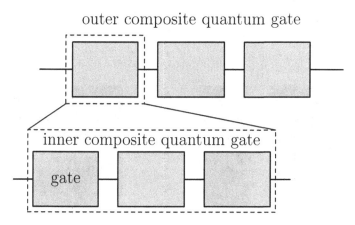

Fig. 2.   Schematic diagram of outer composite quantum gate and inner composite quantum gate.

The CinBB needs $3 \times 4 = 12$ elementary gates because CORPSE and BB1 consist three and four elementary gates, respectively. By taking into account a Residual-Error-Preserving (REP), the number of elementary gates of CinBB can be reduced.[16] The reduced CinBB consists six elementary gates with parameters

$$\theta_1 = \theta_3 = \pi, \quad \theta_2 = 2\pi, \quad \theta_4 = \theta_6 + 2\pi = 2\pi + \theta/2 - k,$$
$$\theta_5 = 2\pi - 2k, \quad \phi_1 = \phi_3 = \phi + \arccos[-\theta/(4\pi)], \qquad (8)$$
$$\phi_2 = 3\phi_1 - 2\phi, \quad \phi_4 = \phi_5 - \pi = \phi_6 = \phi.$$

Density plots of the gate fidelity F of the target elementary gate, BB1, CORPSE, CinBB, and reduced CinBB are shown in Figure 3. These figures show traditional composite quantum gates (BB1 and CORPSE) are robust against either PLE or ORE, and CCCP (CinBB and reduced CinBB) are robust against PLE and ORE simultaneously.

In summary, there are many composite quantum gates in NMR quantum computation to tackle PLE and ORE. Traditional composite quantum gates

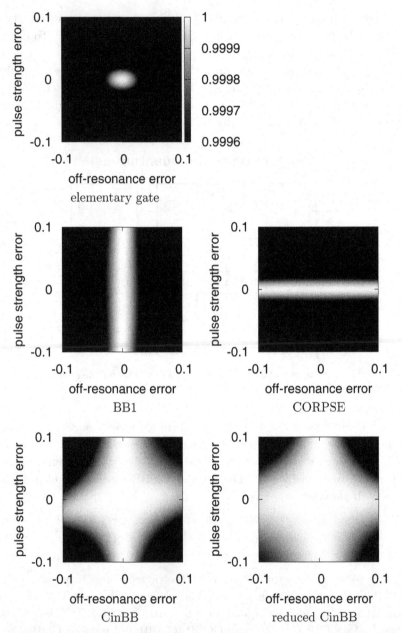

Fig. 3. Density plots of the gate fidelity F of a target elementary gate, BB1, CORPSE, CinBB, and reduced CinBB as a function of the error strengths ε for PLE and f for ORE. The target elementary gate is $R(\pi, 0)$. Traditional composite quantum gates (BB1 and CORPSE) are robust against either PLE or ORE, and the CinBB and the reduced CinBB are robust against PLE and ORE simultaneously.

are robust against either PLE or ORE, and on the other hand concatenated composite pulses (CCCPs) are robust against these two types of errors simultaneously.

Acknowledgments

Y.K. and M.N. would like to thank partial support of a Grant-in-Aid for Scientific Research from the Japan Society for the Promotion of Science (Grant No. 23540470, 25400422).

References

1. J. A. Jones, Prog. Nucl. Magn. Reson. Spectrosc. **59**, 91 (2011).
2. T. D. W. Claridge, High-Resolution NMR Techniques in Organic Chemistry (Elsevier, Amsterdam, 1999).
3. M. H. Levitt, Prog. Nucl. Magn. Reson. Spectrosc. **18**, 61 (1986).
4. J. A. Jones, Philos. Trans. R. Soc. London, Ser. A **361**, 1429 (2003).
5. H. K. Cummins, G. Llewellyn, and J. A. Jones, Phys. Rev. A **67**, 042308 (2003).
6. S. Wimperis, J. Magn. Resonance A **109**, 221 (1994).
7. T. Ichikawa, M. Bando, Y. Kondo, and M. Nakahara, Philos. Trans. R. Soc. A **370**, 4671 (2012).
8. Y. Kondo and M. Bando, J. Phys. Soc. Jpn. **80**, 054002 (2011).
9. P. Zanardi and M. Rasetti, Phys. Lett. A **264**, 94 (1999).
10. Y. Ota and Y. Kondo, Phys. Rev. A **80**, 024302 (2009).
11. M. V. Berry, Proc. R. Soc. London, Ser. A **392**, 45 (1984).
12. F. Wilczek and A. Zee, Phys. Rev. Lett. **52**, 2111 (1984).
13. Y. Aharonov and J. Anandan, Phys. Rev. Lett. **58**, 1593 (1987).
14. M. Nakahara, *Geometry, Topology and Physics* 2nd ed. (Taylor & Francis, Boca Raton, FL, 2003).
15. T. Ichikawa, M. Bando, Y. Kondo, and M. Nakahara, Phys. Rev. A **84**, 062311 (2011).
16. M. Bando, T. Ichikawa, Y. Kondo, and M. Nakahara, J. Phys. Soc. Jpn. **82**, 014004 (2013).

NEW FORMULATION OF STATISTICAL MECHANICS USING THERMAL PURE QUANTUM STATES

SHO SUGIURA* AND AKIRA SHIMIZU

*Department of Basic Science, University of Tokyo, 3-8-1 Komaba, Meguro, Tokyo
153-8902, Japan
* E-mail: sugiura@asone.c.u-tokyo.ac.jp*

We formulate statistical mechanics based on a pure quantum state, which we call a "thermal pure quantum (TPQ) state". A single TPQ state gives not only equilibrium values of mechanical variables, such as magnetization and correlation functions, but also those of genuine thermodynamic variables and thermodynamic functions, such as entropy and free energy. Among many possible TPQ states, we discuss the canonical TPQ state, the TPQ state whose temperature is specified. In the TPQ formulation of statistical mechanics, thermal fluctuations are completely included in quantum-mechanical fluctuations. As a consequence, TPQ states have much larger quantum entanglement than the equilibrium density operators of the ensemble formulation. We also show that the TPQ formulation is very useful in practical computations, by applying the formulation to a frustrated two-dimensional quantum spin system.

Keywords: statistical mechanics, pure quantum state

1. Introduction

In quantum statistical mechanics, equilibrium states are conventionally described by mixed quantum states. By contrast, recent studies have shown the following fact.[1–5] Suppose that one prepares a pure quantum state as superposition of the energy eigenstates whose energies lie in the energy shell $[U - \Delta U, U + \Delta U]$ (U: energy, ΔU: energy width of $o(N)$). Then, almost every such pure state (measured by the Haar measure) gives the expectation values which are equal to those obtained from the microcanonical ensemble average with an exponentially small error, for any "mechanical variables" (See Sec. 2) such as magnetization and the correlation function. This result shows that a pure quantum state can represent a thermal equilibrium state. Motivated by this discovery, we generally call pure quantum states that give the correct equilibrium value for every mechanical variable thermal pure quantum (TPQ) states.[6,7]

However, "genuine thermodynamic variable" such as temperature and the thermodynamic functions cannot be calculated as the expectation values of quantum-mechanical observables. In the ensemble formulation, they are related to the number of states. Therefore, one might think it impossible to obtain genuine thermodynamic variables from a *single* TPQ state. In this paper, however, we will show that genuine thermodynamic variables are related to the normalization constants of appropriate TPQ states. We present one example of such appropriate states, which we call the canonical TPQ state.[7]

While the TPQ state of the previous works is specified by energy, the canonical TPQ state is specified not by energy but by temperature. We will show that the normalization constant of the canonical TPQ state gives the free energy. We also present another TPQ state specified by energy, whose normalization constant gives entropy. We call it the microcanonical TPQ state.[6] We show that the canonical TPQ state can be constructed efficiently from the microcanonical TPQ states.

These results establish a new formulation of statistical mechanics, which enables one to obtain all quantities of statistical-mechanical interest from a *single* realization of a TPQ state. This formulation is not only interesting as fundamental physics but also advantageous in practical applications because one needs only to construct a single pure state by just multiplying the Hamiltonian matrix to a random vector.

2. Canonical TPQ State

We consider a quantum system composed of N sites (or particles). We assume that the dimension D of its Hilbert subspace is finite. [For particle systems, D may be made finite by an appropriate truncation.] We also assume that for this system the ensemble formulation gives correct results, which are consistent with thermodynamics in the thermodynamic limit, $N \to \infty$. Here, we use the term "thermodynamics" in the sense of Refs. 8,9. [This means, for example, that the entropy function is concave.] To exclude foolish operators such as $N^N \hat{H}$, we also assume that every mechanical variable is normalized as $\|\hat{A}\| \leq K N^m$ where m is a constant of $o(N)$ and K is a constant independent of $\hat{A}$ and N. We use quantities per site, e.g., $u \equiv E/N$ and $\hat{h} \equiv \hat{H}/N$. The spectrum of $\hat{h}$ is assumed to be bounded, i.e., $e_{\min} \leq u \leq e_{\max}$.

The canonical TPQ state $|\beta, N\rangle$ is specified by the inverse temperature

β and N (and possibly other variables such as magnetization, on which we do not explicitly write the dependence). In order to generate it, take a random vector

$$|\psi_0\rangle \equiv \sum_i c_i |i\rangle \tag{1}$$

from the whole Hilbert space. Here, $\{|i\rangle\}_i$ is an arbitrary orthonormal basis set of the whole Hilbert space and $\{c_i\}_i$ is a set of random complex numbers drawn uniformly from the $2D$ dimensional sphere, $\sum_i |c_i|^2 = D$. Then, the canonical TPQ state is given by

$$|\beta, N\rangle \equiv \exp\left[\frac{-N\beta\hat{h}}{2}\right] |\psi_0\rangle. \tag{2}$$

As we will see in the next two sections, it correctly gives both the thermodynamic functions and the equilibrium values of the mechanical variables.

We notice that TPQ states are not the "purification" of mixed states (for details of purification, see Ref. 10), because TPQ states are pure states in the D-dimensional Hilbert subspace, i.e., they do not require an ancilla.

3. Thermodynamic Functions and Genuine Thermodynamic Variables

The free energy, which is one of thermodynamic functions, is obtained from the normalization constant of $|\beta, N\rangle$ as

$$f(\beta; N) = \frac{1}{\beta} \ln\langle\beta, N|\beta, N\rangle, \tag{3}$$

where $f(\beta; N) \equiv -(1/\beta N) \ln Z(\beta, N)$ is the free energy density [$Z(\beta, N)$ is the partition function]. Here, we write $(\beta; N)$ instead of (β, N) in order to indicate that $f(\beta; N)$ converges to the N-independent one, $f(\beta)$.

Using the random matrix theory and the generalized Markov inequality, the error probability is evaluated as

$$\mathrm{P}\left(\left|\frac{\langle\beta, N|\beta, N\rangle}{\exp[-N\beta f(1/\beta; N)]} - 1\right| \geq \epsilon\right)$$
$$\leq \frac{1}{\epsilon^2 \exp[2N\beta\{f(1/2\beta; N) - f(1/\beta; N)\}]}, \tag{4}$$

where $\mathrm{P}(\cdot)$ is the probability that an event $\cdot$ happens. Since $f(1/2\beta; N) - f(1/\beta; N)$ is positive and $\Theta(1)$[11] from thermodynamics,[9] the r.h.s. of inequality (4) is $\Theta(1/\epsilon^2 \exp[N])$. Therefore, a single realization of the canonical TPQ state almost always gives the correct thermodynamic function with an exponentially small error. In another word,

$$\frac{1}{\beta} \ln\langle \beta, N | \beta, N \rangle \xrightarrow{P} f(\beta) \tag{5}$$

where $\xrightarrow{P}$ denotes convergence in probability.

All genuine thermodynamic variables and any other thermodynamic functions can be obtained from $f(\beta)$ by differentiation and the Legendre transformation.

4. Mechanical Variables

In the previous section, we have shown that the canonical TPQ state correctly gives the free energy. The equilibrium values of all macroscopic quantities are derived from derivatives of the free energy. For mechanical variables, one can also obtain their equilibrium values as the expectation values in the TPQ state.

The expectation value of a mechanical variable $\hat{A}$ in the canonical TPQ state

$$\langle \hat{A} \rangle_{\beta,N}^{\text{TPQ}} \equiv \frac{\langle \beta, N | \hat{A} | \beta, N \rangle}{\langle \beta, N | \beta, N \rangle} \tag{6}$$

gives the equilibrium value with an exponentially small error. Like the ensemble average, the expectation value is useful in many practical applications.

The squared average of the difference between this expectation value and the canonical ensemble average

$$\langle \hat{A} \rangle_{\beta,N}^{\text{ens}} \equiv \frac{\text{Tr}\left[e^{-N\beta\hat{h}}\hat{A}\right]}{Z(\beta, N)} \tag{7}$$

is estimated as

$$\overline{(\langle \hat{A} \rangle_{\beta,N}^{\text{TPQ}} - \langle \hat{A} \rangle_{\beta,N}^{\text{ens}})^2} \leq \frac{\langle (\Delta\hat{A})^2 \rangle_{2\beta,N}^{\text{ens}} + (\langle A \rangle_{2\beta,N}^{\text{ens}} - \langle A \rangle_{\beta,N}^{\text{ens}})^2}{\exp[2N\beta\{f(1/2\beta; N) - f(1/\beta; N)\}]}, \tag{8}$$

where $\langle (\Delta\hat{A})^2 \rangle_{\beta,N}^{\text{ens}} \equiv \langle (\hat{A} - \langle A \rangle_{\beta,N}^{\text{ens}})^2 \rangle_{\beta,N}^{\text{ens}}$. Using the generalized Markov inequality, we get an upper bound of the error probability as

$$P\left(\left|\langle \hat{A} \rangle_{\beta,N}^{\text{TPQ}} - \langle \hat{A} \rangle_{\beta,N}^{\text{ens}}\right| \geq \epsilon\right) \leq \frac{1}{\epsilon^2} \frac{\langle (\Delta\hat{A})^2 \rangle_{2\beta,N}^{\text{ens}} + (\langle A \rangle_{2\beta,N}^{\text{ens}} - \langle A \rangle_{\beta,N}^{\text{ens}})^2}{\exp[2N\beta\{f(1/2\beta; N) - f(1/\beta; N)\}]}. \tag{9}$$

Since $\|\hat{A}\| < KN^m$ (Sec. 2), the r.h.s. is $\Theta(N^m/\epsilon^2 \exp[N])$. Therefore, a single realization of the canonical TPQ state almost always gives the correct equilibrium values of any mechanical variables with an exponentially small error.

We have shown that the equilibrium values of both mechanical and genuine thermodynamic variables are obtained from a single realization of the TPQ state. In this sense, we have established a new formulation of statistical mechanics based on a pure quantum state.

5. A Numerical Application

Since our formulation requires only a single pure state for each equilibrium state, it is a powerful tool for practical applications. To illustrate this fact, we apply our formulation to a numerical computation in this section.

We present the result for spin-1/2 Kagome lattice Heisenberg antiferromagnet (KHA). This system is known to be hard to analyze because of frustration. On the ground of the numerical diagonalization of small clusters up to N=18, it was suggested that the specific heat of KHA would have double peaks at low temperature.[12-15]

In Fig. 1, we show our results for the specific heat. [Some detail of the computation will be described in Sec. 7.] The results for $N = 18a$ and b correspond to different shapes of the clusters. These results agree well with the previous results calculated by the numerical diagonalization, and show the double peaks. However, the peak at lower temperature vanishes for larger sizes, $N = 27$ and 30 [which cannot be treated by the numerical diagonalization]. We have obtained the results for these two clusters from a single realization of the canonical TPQ state. This suggests that the peak at lower temperature would be absent in the thermodynamic limit.

Another important result is the entropy density and the free energy density, shown in Fig. 2. We observe that there remains 45% of the total entropy $(= N \ln 2)$ at $T = 0.2J$. This is a consequence of strong frustration, which makes this system hard to analyze.

We emphasize again that these variables for $N = 27$ and 30 have been obtained from a single realization of the canonical TPQ state. Moreover, recalling inequality (9), we can estimate the probabilistic error of the result of the specific heat by using the result of the free energy density. The error is estimated to be less than 1% down to $T = 0.1J$. Thus, our new results for $N = 27$ and 30 are reliable enough for most purposes.

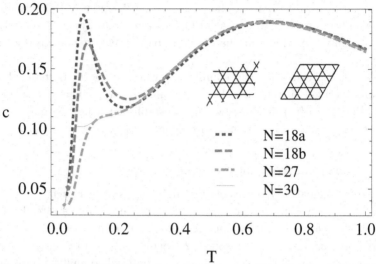

Fig. 1. c vs. T of the KHA. The shapes of clusters of $N = 30$, 27 and 18a, 18b are shown in the right, left and in Ref. 13, respectively.

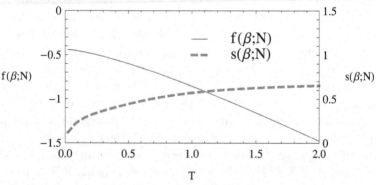

Fig. 2. f and s vs. T for $N = 30$. The shape of cluster is shown in Fig. 1.

6. Microcanonical TPQ State

While we have generally defined TPQ states roughly in Sec. 1, we define it rigorously as follows. When a state $|\psi\rangle$ is generated from some probability measure, it is called a TPQ state if

$$\langle \hat{A} \rangle_N^{\psi} \overset{P}{\to} \langle \hat{A} \rangle_N^{\text{ens}} \tag{10}$$

uniformly for every mechanical variable $\hat{A}$ as $N \to \infty$. Here, $\langle \hat{A} \rangle_N^\psi \equiv \langle \psi | \hat{A} | \psi \rangle / \langle \psi | \psi \rangle$, $\langle \cdot \rangle_N^{\text{ens}}$ is the ensemble average, and '$\xrightarrow{P}$' denotes convergence in probability.

This definition clearly shows that a single realization of the TPQ state for sufficiently large N is enough to evaluate the equilibrium values of all *mechanical* variables. Among such TPQ states are the random state in the energy shell[1-5] and the canonical TPQ state. The latter has an additional special property that it also gives the equilibrium values of all *genuine thermodynamic* variables. In this section, we present another TPQ state, called the microcanonical TPQ state, which also has this special property.

Starting from the random vector $|\psi_0\rangle$ given by Eq. (1), the microcanonical TPQ state is defined by

$$|k\rangle \equiv (l - \hat{h})^k |\psi_0\rangle \quad (k = 0, 1, 2, \cdots), \tag{11}$$

where l is an arbitrary constant s.t. $l \geq \{\text{maximum eigenvalue of } \hat{h}\}$. The equilibrium value of the energy density is obtained by

$$\frac{\langle k | \hat{h} | k \rangle}{\langle k | k \rangle} \equiv u_k. \tag{12}$$

More generally, the equilibrium value of a mechanical variable $\hat{A}$ is obtained by

$$\frac{\langle k | \hat{A} | k \rangle}{\langle k | k \rangle}. \tag{13}$$

We can show that this value, with increasing N, approaches the expectation value for the microcanonical ensemble of energy $N u_k$. Thus, $|k\rangle$ satisfies the above condition for a TPQ state. Furthermore, it gives the entropy density $s(u)$ as

$$\frac{1}{N} \ln \langle k | k \rangle - \frac{2k}{N} \ln(l - u_k) \xrightarrow{P} s(u_k). \tag{14}$$

Since the microcanonical TPQ state is generated by multiplying the polynomial of $\hat{h}$ to the random vector $|\psi_0\rangle$, it can be generated easily, e.g., in a computer.

7. Expansion of the Canonical TPQ State

The canonical TPQ state can be decomposed as the superposition of the microcanonical TPQ states. This decomposition enables one to perform numerical calculation efficiently.

We apply simple Taylor expansion to $\exp[N\beta(l - \hat{h})/2]$ as

$$|\beta, N\rangle = e^{-N\beta l/2} \sum_{k=0}^{\infty} \frac{(N\beta/2)^k}{k!} |k\rangle \tag{15}$$

$$= e^{-N\beta l/2} \sum_{k=0}^{\infty} R_k |\psi_k\rangle \tag{16}$$

where $|\psi_k\rangle \equiv |k\rangle/\sqrt{\langle k|k\rangle}$ is the normalized microcanonical TPQ state, and $R_k \equiv \sqrt{\langle k|k\rangle}(N\beta/2)^k/k!$.

Although this Taylor expansion is the sum of infinite terms, relevant k's are not so many. To see the contribution of each $|k\rangle$ to $|\beta, N\rangle$, we focus on R_k (> 0). We can show that R_k takes the maximum value for k such that u_k is closest to $\langle \beta, N|\hat{h}|\beta, N\rangle$. The values of R_k for other k's decay exponentially fast as the corresponding u_k gets further from $\langle \beta, N|\hat{h}|\beta, N\rangle$. Thus, we can efficiently generate the canonical TPQ state from a small number of the microcanonical TPQ states, which can be numerically generated easily.

The numerical results shown in Sec. 5 have been calculated using the above relation.

8. Quantum and Thermal Fluctuations

To better understand the TPQ states, we now discuss the "quantum fluctuation" and "thermal fluctuation". For concreteness, we consider the canonical TPQ state $|\beta, N\rangle$ and the canonical density operator $\hat{\rho} = e^{-\beta N\hat{h}}/Z$.

In the ensemble formulation, it is often said that a fluctuation of a mechanical variable $\langle(\Delta\hat{A})^2\rangle^{\text{ens}} \equiv \langle(\hat{A} - \langle\hat{A}\rangle^{\text{ens}})^2\rangle^{\text{ens}}$ can be decomposed into the quantum fluctuation $\langle(\Delta\hat{A})^2\rangle_{\text{q}}^{\text{ens}}$ and the thermal one $\langle(\Delta\hat{A})^2\rangle_{\text{t}}^{\text{ens}}$, i.e.,

$$\langle(\Delta\hat{A})^2\rangle^{\text{ens}} = \langle(\Delta\hat{A})^2\rangle_{\text{q}}^{\text{ens}} + \langle(\Delta\hat{A})^2\rangle_{\text{t}}^{\text{ens}}. \tag{17}$$

The thermal fluctuation, whose specific expression will be given below, is conventionally interpreted as a result of mixing many quantum states to form $\hat{\rho}$,

$$\hat{\rho} = \sum_n (e^{-\beta N e_n}/Z)|n\rangle\langle n|, \tag{18}$$

where e_n and $|n\rangle$ are eigenvalue and eigenstate, respectively, of $\hat{h}$. Consequently, it is conventionally concluded that the thermal fluctuation of most mechanical variables does not vanish at any finite temperature.

In the TPQ formulation, by contrast, $|\beta, N\rangle$ is a pure quantum state and therefore does not have such "thermal fluctuation", i.e., $\langle(\Delta\hat{A})^2\rangle_t^{\text{TPQ}} = 0$ at all temperature. The TPQ state has only the quantum fluctuation, i.e.,

$$\langle(\Delta\hat{A})^2\rangle^{\text{TPQ}} = \langle(\Delta\hat{A})^2\rangle_q^{\text{TPQ}} \equiv \langle(\hat{A} - \langle\hat{A}\rangle^{\text{TPQ}})^2\rangle^{\text{TPQ}}. \tag{19}$$

In other words, all fluctuations are included in the quantum fluctuation.

We have thus found that $\hat{\rho}$ and $|\beta, N\rangle$, which represent the same equilibrium state, give different values of the quantum and thermal fluctuations. This does not lead to any contradiction in experimentally-observable quantities because

$$\langle(\Delta\hat{A})^2\rangle^{\text{ens}} = \langle(\Delta\hat{A})^2\rangle^{\text{TPQ}}, \tag{20}$$

which are the only observable quantities in the above discussion. The quantum and thermal fluctuations, $\langle(\Delta\hat{A})^2\rangle_q^{\text{ens}}$ and $\langle(\Delta\hat{A})^2\rangle_t^{\text{ens}}$, are, separately, not observable quantities. To see this, let us write them down explicitly. We note that ρ has the following form,

$$\hat{\rho} \equiv \sum_\lambda w_\lambda |\lambda\rangle\langle\lambda|, \tag{21}$$

where $\{w_\lambda\}_\lambda$ is a set of positive numbers such that $\sum_\lambda w_\lambda = 1$, and $\{|\lambda\rangle\}_\lambda$ is some set of states (which is $\{|n\rangle\}_n$ in Eq. (18)). In general, $\hat{A}$ fluctuates quantum-mechanically in each state $|\lambda\rangle$. Hence, it may be reasonable to define $\langle(\Delta\hat{A})^2\rangle_q^{\text{ens}}$ as the average of the fluctuation $\langle\lambda|(\hat{A} - \langle\lambda|\hat{A}|\lambda\rangle)^2|\lambda\rangle$ over $|\lambda\rangle$'s, i.e.,

$$\langle(\Delta\hat{A})^2\rangle_q^{\text{ens}} \equiv \sum_\lambda w_\lambda \langle\lambda|(\hat{A} - \langle\lambda|\hat{A}|\lambda\rangle)^2|\lambda\rangle. \tag{22}$$

This and Eq. (17) yield the thermal fluctuation as

$$\langle(\Delta\hat{A})^2\rangle_t^{\text{ens}} = \sum_\lambda w_\lambda \langle\lambda|\hat{A}|\lambda\rangle^2 - \left(\sum_\lambda w_\lambda \langle\lambda|\hat{A}|\lambda\rangle\right)^2. \tag{23}$$

If we take $w_\lambda = e^{-\beta N e_n}/Z$ and $|\lambda\rangle = |n\rangle$, we find that $\langle(\Delta\hat{A})^2\rangle_t^{\text{ens}} > 0$ for most mechanical variables at finite temperature.

However, it is well-known that $|\lambda\rangle$'s in Eq. (21) need not be orthogonal to each other.[10] As a result, there are infinitely many possible choices of $\{|\lambda\rangle\}_\lambda$ and $\{w_\lambda\}_\lambda$ for the same $\hat{\rho}$.[10] The experimentally-observable fluctuation $\langle(\Delta\hat{A})^2\rangle^{\text{ens}}$ is invariant under the change of $\{w_\lambda\}_\lambda$ and $\{|\lambda\rangle\}_\lambda$. By contrast, both $\langle(\Delta\hat{A})^2\rangle_q^{\text{ens}}$ and $\langle(\Delta\hat{A})^2\rangle_t^{\text{ens}}$ do alter under the change of $\{w_\lambda\}_\lambda$ and $\{|\lambda\rangle\}_\lambda$. This fact clearly shows that the quantum and thermal fluctuations are, separately, not experimentally-observable quantities. In other words, they are, separately, metaphysical quantities.

It is instructive to consider a classical mixture

$$\hat{\rho}' \equiv \frac{1}{R} \sum_{r=1}^{R} \frac{|\beta, N, r\rangle\langle\beta, N, r|}{\langle\beta, N, r|\beta, N, r\rangle} \tag{24}$$

of many realizations $|\beta, N, 1\rangle, |\beta, N, 2\rangle, \cdots, |\beta, N, R\rangle$ of the canonical TPQ state. Since each $|\beta, N, r\rangle$ represents the same equilibrium state, so does $\hat{\rho}'$. If we define the quantum and thermal fluctuations in $\hat{\rho}'$ in the same way as Eqs. (22) and (23), we find that the thermal fluctuation is exponentially small for all mechanical variables. This shows that mixing many states does not necessarily give "thermal fluctuation". Since the thermal fluctuation in $\hat{\rho}'$ is negligible, we do not need to take an average over many relizations, but only need to pick up a single realization.

9. Entanglement

We have shown that the TPQ states and the density operators of the statistical ensembles give identical results for all quantities of statistical-mechanical interest. That is, as far as one looks at macroscopic quantities, one cannot distinguish between these states. However, the TPQ states are pure quantum states while the density operators in the ensemble formulation (i.e., the Gibbs states) are mixed states. Therefore, the situation changes when we look at entanglement. We discuss this point by studying entangelement of the microcanonical TPQ state.

To investigate entangelement of the TPQ state, we study its reduced density operator ρ_q that is obtained by tracing out $N - q$ sites. Its purity is defined by $\text{Tr}(\rho_q^2)$. Since a TPQ state is a pure quantum state, this purity is a good measure of its entanglement. The smaller the purity is, the more entanglement the TPQ state has.

In Fig. 3, we plot the minimum value of the purity (triangles ▲) and the average value of the purity of the random vector $|\psi_0\rangle$ (inverse triangles ▼). It is seen that $|\psi_0\rangle$ has almost maximum (exponentially large) entanglement.[16] The lines are the purity of the microcanonical TPQ states with different values of the energy density. It is seen that the TPQ states have exponentailly large entanglement, and that the entanglement gets larger at higher energy, i.e., at higher temperature. This result is in marked contrast to entanglement of the density operator of the ensemble formulation, because the latter has less entanglement at higher temperature. [For example, with increasing temperature the canonical density operator approaches identity, which has no entanglement in any reasonable entanglement measure.]

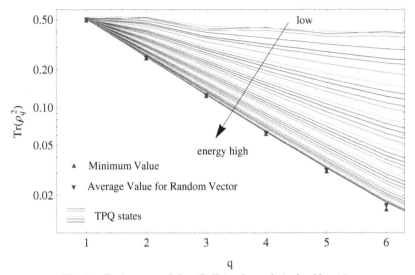

Fig. 3. Purity vs. q of the 1D Heisenberg chain for $N = 16$.

However, this is not a contradiction but a natural consequence of the nature of entanglement. The purity of ρ_q is related to N-body correlation functions of the TPQ state. Such higher-order correlation functions represent microscopic details of the TPQ state. Therefore, the great difference in entanglement between the TPQ states and the Gibbs states indicates a great difference in microscopic details. It is not surprising that such *microscopically* completely different states give identical results for *macroscopic* quantities, and thus represent the same equilibrium state.

10. Conclusion

In this paper, we have established a new formulation of statistical mechanics based on new TPQ states. A single realization of the TPQ state gives equilibrium values of all mechanical and genuine thermodynamic variables and thermodynamic functions, with an exponentially small error. However, the TPQ states are completely different from the Gibbs states of the ensemble formulation. We have illustrated this fact by showing great difference of entanglement between them. There are many possible TPQ states, such as the canonical TPQ state and the microcanonical one. The canonical TPQ state can be generated from the microcanonical ones, and the microcanonical ones can be obtained easily in a computer. This fact makes the TPQ formulation advantageous in practical applications.

References

1. A. Sugita, RIMS Kokyuroku (Kyoto) **1507**, 147 (2006) [in Japanese].
2. A. Sugita, Nonlinear Phenom. Complex Syst. **10**, 192 (2007).
3. S. Popescu, A.J. Short, and A. Winter, Nature Phys. **2**, 754 (2006).
4. S. Goldstein *et al*, Phys. Rev. Lett. **96**, 050403 (2006).
5. P. Reimann, Phys. Rev. Lett. **99**, 160404 (2007).
6. S. Sugiura and A. Shimizu, Phys. Rev. Lett. **108**, 240401 (2012).
7. S. Sugiura and A. Shimizu, Phys. Rev. Lett. **111**, 010401 (2013).
8. H. B. Callen, *Thermodynamics* (John Wiley and Sons, New York, 1960).
9. A. Shimizu, *Netsurikigaku no Kiso* (Principles of Thermodynamics) (University of Tokyo Press, Tokyo, 2007) [in Japanese].
10. M. A. Nielsen and I. L. Chuang: *Quantum Computation and Quantum Information* (Cambridge University Press, Cambridge, 2000).
11. For $f(N), g(N) > 0$, $f = \Theta(g)$ denotes that $\exists a > 0, \exists b > 0, \exists N_0$ such that $ag(N) \leq f(N) \leq bg(N)$ for all $N > N_0$.
12. V. Elser, Phys. Rev. Lett. **62**, 2405 (1989).
13. N. Elstner and A. P. Young, Phys. Rev. B **50**, 6871 (1994).
14. P. Sindzingre, G. Misguich, C. Lhuillier, B. Bernu, L. Pierre, Ch. Waldtmann, and H.-U. Everts, Phys. Rev. Lett. **84**, 2953 (2000).
15. M. Isoda, H. Nakano, and T. Sakai, J. Phys. Soc. Jpn **80**, 084704 (2011).
16. A. Sugita and A. Shimizu, J. Phys. Soc. Jpn. **74**, 1883 (2005).

THERMODYNAMICS IN UNITARY TIME EVOLUTION

TATSUHIKO N. IKEDA

Department of Physics, University of Tokyo,
Tokyo 113-0033, Japan
E-mail: ikeda@cat.phys.s.u-tokyo.ac.jp
http://cat.phys.s.u-tokyo.ac.jp/~ikeda/index.html

A brief review is provided on the recent developments on the theory of thermalization under unitary time evolution in isolated quantum systems. In particular, we focus on the theory of equilibration, which states the existence of the effective stationary state, and the eigenstate thermalization hypothesis, which is a scenario for the stationary state being regarded as the microcanonical ensemble. Then, we introduce other possible scenarios and compare them from the viewpoint of the finite-size scaling. We also discuss the relative importance among these scenarios.

Keywords: Thermalization; Isolated Quantum Systems; Foundation of Statistical Mechanics.

1. Introduction

An isolated quantum system is an ideal system not only for quantum mechanics but also for quantum thermodynamics and statistical mechanics. However, how the latter two are validated from the former is a quite nontrivial question. Quantum mechanics states that a quantum state is described by a pure state $|\psi\rangle$ and that the time evolution operator is unitary, $U(t) = e^{-iHt}$, where H is the Hamiltonian of the system and the Planck constant is set to be unity. The state at time t, $|\psi(t)\rangle = U(t)|\psi\rangle$, remains pure and time-dependent. On the other hand, thermodynamics presupposes that the system reaches equilibrium as time goes on and statistical mechanics does that the equilibrated state is described by the microcanonical ensemble $\rho_{E,\delta}^{\mathrm{mic}} = \mathcal{N}_{E,\delta}^{-1} \sum_{n \, \mathrm{s.t.} |E_n - E| \leq \delta} |E_n\rangle \langle E_n|$, where $|E_n\rangle$ is an eigesnstate of H with eigenenergy E_n, E the total energy, δ a sub-extensive energy width, and $\mathcal{N}_{E,\delta}$ the number of the eigenstates whose eigenenergies lie between $E - \delta$ and $E + \delta$. The microcanonical ensemble is a mixed state and time-independent.

The above problem concerning the foundation of statistical mechanics has recently attracted revived attention due to the experimental realizations of isolated quantum systems such as ultracold atoms and ions.[1] Since these systems are trapped under high vacuum and have little interaction with particles outside the system, they can be regarded as isolated systems.

This paper is organized as follows. First, we briefly review some basic concepts especially equilibration,[2] which states how an effective stationary state emerges from the time-dependent pure state $|\psi(t)\rangle$. Second, we see a scenario, which is called the eigenstate thermalization hypothesis (ETH), for the stationary state being regarded as the microcanonical ensemble.[3] Third, two other possible scenarios, the eigenstate randomization hypothesis (ERH)[14] and the typicality argument for special initial conditions. Fourth, we quantitatively discuss the relative importance among these scenarios with respect to the finite-size scaling analysis.[15]

2. Basic Concepts

2.1. *Typicality*

Before discussing the question raised in the previous section, we first ask the following question: can a single pure quantum state $|\psi\rangle$ be regarded as the microcanonical ensemble? According to the fact called the typicality, the answer is yes.

The crucial idea is restricting the class of observables of interest. The condition that the density matrices, $|\psi\rangle\langle\psi|$ and $\rho_{E,\delta}^{\mathrm{mic}}$, are equal means that $\langle\psi|A|\psi\rangle = \mathrm{tr}(\rho_{E,\delta}^{\mathrm{mic}}A)$ holds for any Hermitian operator A. This set of Hermitian operators involves, for example, the correlation of all the particles of the system. However, we do not and cannot check if the expectation values of such operators are described by $\rho_{E,\delta}^{\mathrm{mic}}$ when we state that the system is at equilibrium. Thus, we can say that a single pure state $|\psi\rangle$ is at equilibrium if $\langle\psi|A|\psi\rangle = \mathrm{tr}(\rho_{E,\delta}^{\mathrm{mic}}A)$ holds for every local or few-body observable A.

The typicality states that almost all pure states are at equilibrium in the above sense. Let us elaborate this statement in the following. We define a subspace $\mathcal{H}_{E,\delta} = \mathrm{span}\{|E_n\rangle \mid |E_n - E| \leq \delta\}$, where the right-hand side denotes the subspace spanned by the eigenstates with eigenenergy between $E - \delta$ and $E + \delta$. An arbitrary pure state in $\mathcal{H}_{E,\delta}$ is expanded in terms of the eigenstates as

$$|\psi\rangle = \sum_n{}' c_n |E_n\rangle, \qquad (1)$$

where $\sum_n{}'$ denotes the sum over n such that $|E_n - E| \leq \delta$ and c_n's are complex numbers, which satisfy the normalization condition: $\sum_n {}'|c_n|^2 = 1$. In other words, each pure state in $\mathcal{H}_{E,\delta}$ is mapped to a point on $(2d-1)$-dimensional unit sphere, S^{2d-1}, where $d \equiv \dim \mathcal{H}_{E,\delta}$[a]. Now, we define the probability distribution generating pure states as the uniform distribution on S^{2d-1}. This is nothing but the uniform Haar measure on $\mathcal{H}_{E,\delta}$. The average of each density matrix $|\psi\rangle\langle\psi|$ over the uniform measure is obviously $\rho_{E,\delta}^{\mathrm{mic}}$ although the variance of $|\psi\rangle\langle\psi|$ is large. Now we consider the reduced density matrix onto a small subsystem S. Since this reduction is linear, the average of $\mathrm{tr}_{\overline{S}}\,|\psi\rangle\langle\psi|$ is $\mathrm{tr}_{\overline{S}}\rho_{E,\delta}^{\mathrm{mic}}$, where $\mathrm{tr}_{\overline{S}}$ denotes the partial trace over the rest of the subsystem S. The typicality shows that the average distance between $\mathrm{tr}_{\overline{S}}\,|\psi\rangle\langle\psi|$ and $\mathrm{tr}_{\overline{S}}\rho_{E,\delta}^{\mathrm{mic}}$ is very small when the subsystem is much smaller than the whole system. This implies that almost all pure states in the whole system are at equilibrium if we look into small subsystems.

This striking result has some implications on the formulation of statistical mechanics. Statistical mechanics presupposes the microcanonical ensemble, which is equivalent to the average over the uniform sampling on $\mathcal{H}_{E,\delta}$. However, the typicality states that a single typical sample from the uniform measure on $\mathcal{H}_{E,\delta}$ is equivalent to the average, and thus, taking average is not necessary any longer. A new formulation of statistical mechanics and a new method for numerical calculations of problems in statistical mechanics have been studied.[4,5]

2.2. *Equilibration*

2.2.1. *Motivation*

The typicality is not enough to explain how a pure state reaches equilibrium in unitary time evolution although it suggests that it is possible in principle. It is because the uniform Haar measure on $\mathcal{H}_{E,\delta}$ is not achieved in unitary time evolution. Let us pick up a pure state in the form of Eq. (1) and regard it as the initial state at time $t = 0$. Then, the quantum state at time t is represented as

$$|\psi(t)\rangle = \sum_n {}'c_n(t)\,|E_n\rangle, \tag{2}$$

where $c_n(t) = c_n \mathrm{e}^{-\mathrm{i}E_n t}$. Although we have different realizations of $|\psi(t)\rangle$'s, they do not uniformly spread over $\mathcal{H}_{E,\delta}$ because each weight on the eigen-

---

[a]Here we regarded two pure states which are different only by an overall phase factor as being mapped to two different points although these states are to be identified. However, this makes no problem in the following arguments because all of the pure states are equally multiple-counted.

states is constrained: $|c_n(t)|^2 = |c_n|^2$. Namely, the measure of the set of states generated by unitary evolution is zero in the uniform Haar measure.

Thus, we should ask the following question: is there a state as which $|\psi(t)\rangle$ is regarded at almost all t? If so, it implies that $|\psi(t)\rangle$ becomes stationary in unitary time evolution. This phenomenon is called equilibration and, in fact, happens under two assumptions described below. Here, we stress that equilibration just states the existence of the effective stationary state but does not require that the stationary state is equivalent to the microcanonical ensemble. The latter question is called thermalization and discussed in the following section.

2.2.2. *Assumptions*

For convenience, we assume that the system is for particles or spins. The number of particles or spins denoted by N is assumed to be finite and the volume is also assumed to be finite. These assumptions imply that the spectrum of the Hamiltonian is discrete. Let us pick up an initial state $|\psi\rangle$ and consider its unitary time evolution generated by the Hamiltonian H. The quantum state at time t is expanded in terms of the eigenstates of H as

$$|\psi(t)\rangle = \sum_n c_n e^{-iE_n t} |E_n\rangle, \tag{3}$$

where c_n's are the expansion coefficients of the initial state. We put the following assumptions on the initial state and the Hamiltonian, respectively.[2,6] For convenience, we assume that the spectrum of H has no degeneracy.[b] This assumption can be removed generally.[7]

Assumption 2.1 (Small Inverse Participation Ratio). *The inverse participation ratio (IPR) of the initial state* $|\psi\rangle$ *defined as* $Q \equiv \sum_n |c_n|^4$ *is exponentially small with respect to* N:

$$Q = e^{-O(N)}. \tag{4}$$

We note that the number of the energy eigenstates in a given energy interval is of the order of $e^{O(N)}$. Since Q^{-1} denotes the effective number of the energy eigenstates which are superposed, this assumption states that the superposition is not too localized.

---

[b]Tasaki assumed that H has no degeneracy and then assumed the nonresonance condition. Reimann's assumption is equivalent to the set of these. Short assumed only the nonresonance condition but he called it non-degenerate energy gaps.

Assumption 2.2 (Nonresonance Condition). *If $E_{n_1} - E_{n_2} = E_{n_3} - E_{n_4} \neq 0$, then $E_{n_1} = E_{n_3}$ and $E_{n_2} = E_{n_4}$.*

This assumption states that all the differences between two eigenenergies are different. This does not hold, for example, for free bosons or fermions. However, it is believed to hold for interacting systems, which exhibit the quantum chaos where the energy spectrum is very complex.[2]

2.2.3. *Theory of equilibration*

Picking up an observable A, we consider the time evolution of its expectation value. Although it never converges to a certain value, we can show that the fluctuation of it in the long run is negligibly small. To be more precise, the following inequality holds.

Theorem 2.1. *The variance of the expectation value of an observable A in the long run is bounded from above as*

$$\lim_{T \to \infty} \int_0^T \frac{dt}{T} \left[\langle \psi(t)|A|\psi(t) \rangle - \mathrm{tr}(\rho_{\mathrm{DE}} A) \right]^2 \leq Q \Delta_A^2, \tag{5}$$

where Δ_A is the difference between the maximum and minimum of the eigenvalues of A and $\rho_{\mathrm{DE}} = \sum_n |c_n|^2 |E_n\rangle \langle E_n|$.

Here, we make some remarks on this inequality. First, ρ_{DE} is called the diagonal ensemble because it consists only of the diagonal elements of the instantaneous density matrix, $|\psi(t)\rangle \langle \psi(t)| = \sum_{n,m} c_m c_n^* e^{-i(E_m - E_n)t} |E_m\rangle \langle E_n|$. Second, the expectation value of A over the diagonal ensemble is equal to the long-time average of the time-dependent expectation value, $\mathrm{tr}(\rho_{\mathrm{DE}} A) = \lim_{T \to \infty} \int_0^T \frac{dt}{T} \langle \psi(t)|A|\psi(t) \rangle$, due to the formula, $\lim_{T \to \infty} \int_0^T \frac{dt}{T} e^{i(E_n - E_m)t} = \delta_{mn}$. Third, the upper bound, $Q\Delta_A^2$, is much smaller than the typical magnitude of A if Δ_A grows at most in a power law of N because Q is exponentially small. Thus, equilibration seems to occur as long as we only look at those observables in the sense that $\langle \psi(t)|A|\psi(t) \rangle \sim \mathrm{tr}(\rho_{\mathrm{DE}} A)$ holds at almost every t. Fourth, there remains an infinitesimal portion of t which gives the timings when $\langle \psi(t)|A|\psi(t) \rangle$ significantly deviates from $\mathrm{tr}(\rho_{\mathrm{DE}} A)$. This corresponds to the initial time interval out of equilibrium and the recurrences which cannot be avoided in the finite-size systems.

Intuitively speaking, equilibration occurs due to the phase randomization in the following sense. The expectation value of A at time t has two contributions, $\langle \psi(t)|A|\psi(t) \rangle = \sum_n |c_n|^2 \langle E_n|A|E_n \rangle +$

$\sum_{m \neq n} e^{i(E_n - E_m)t} c_m c_n^* \langle E_n|A|E_m \rangle$. The first term on the right-hand side corresponds to $\text{tr}(\rho_{\text{DE}}A)$. Thus, at almost all t, the terms in $\sum_{m \neq n} e^{i(E_n - E_m)t} c_m c_n^* \langle E_n|A|E_m \rangle$ are summed up to be zero. If the expectation value of A significantly deviated from $\text{tr}(\rho_{\text{DE}}A)$ in the initial state, the phases of the coefficients, c_n's, were finely tuned so that $\sum_{m \neq n} c_m c_n^* \langle E_n|A|E_m \rangle$ gives nonzero. Then, the relative phase goes random and the off-diagonal contribution becomes zero as time goes on.

3. Thermalization

Thermalization is the empirical phenomenon in which expectation values of observables after equilibration coincide with the ones calculated by the microcanonical ensemble $\rho_{E,\delta}^{\text{mic}}$. According to the previous section, the expectation value of A at equilibrium is

$$\text{tr}(\rho_{\text{DE}}A) = \sum_n |c_n|^2 A_n, \tag{6}$$

where $A_n \equiv \langle E_n|A|E_n \rangle$ is the eigenstate expectation value (EEV) of A. On the other hand, the microcanonical ensemble average of A is

$$\text{tr}(\rho_{E,\delta}^{\text{mic}}A) = \mathcal{N}_{E,\delta}^{-1} \sum_{n \, \text{s.t.} \, |E_n - E| \leq \delta} A_n. \tag{7}$$

While Eq. (6) depends on the details of the initial state, $\{|c_n|^2\}_n$, Eq. (7) does only on the total energy E and the energy window δ. In this section, we see three scenarios which explain thermalization. We assume that the support of $\{|c_n|^2\}_n$ and the standard variance of the total energy of $|\psi\rangle$ are as large as δ.

3.1. *Eigenstate Thermalization Hypothesis (ETH)*

The eigenstate thermalization hypothesis (ETH) states that A_n becomes a constant in an energy window in the thermodynamic limit. If the ETH holds, it immediately follows that Eqs. (6) and (7) are equal. In this sense, the ETH is a sufficient condition for thermalization.

The ETH was originally conjectured by Deutsch[8] with the random matrix theory and by Srednicki[9] with semi-classical analyses. Recent developments of performance of computers have enabled us to investigate the ETH by the exact diagonalization of Hamiltonians.[3] But this technique allows us to test the ETH for the system whose dimension of Hilbert space is no more than about 20,000 which corresponds to 5 particles hopping on 21 sites.[3]

The ETH has been tested numerically, with the size of the system fixed, in terms of the characteristics of Hamiltonians, such as integrability,[3] quantum chaos,[10,11] and many-body localization.[12] According to these studies, the ETH seems to be better satisfied for nonintegrable and quantum-chaotic systems than for integrable ones. In the nonintegrable regime, the ETH is less satisfied as the system is more localized.

In spite of these studies, the ETH remains a conjecture because a finite-size scaling analysis has not been done. In the finite-size scaling analysis, the ETH can be stated in the weak and strong senses.[13] The ETH in the weak sense states that the variance of the EEV, A_n,

$$\sigma_A \equiv \left\{ \mathcal{N}_{E,\delta}^{-1} \sum_{n \text{ s.t.} |E_n - E| \leq \delta} \left[A_n - \text{tr}(\rho_{E,\delta}^{\text{mic}} A) \right]^2 \right\}^{1/2} \tag{8}$$

converges to zero in the thermodynamic limit. On the other hand, the ETH in the strong sense states that the maximum deviation of the EEV from its average,

$$\Delta_A' \equiv \max_{n \text{ s.t.} |E_n - E| \leq \delta} \left| A_n - \text{tr}(\rho_{E,\delta}^{\text{mic}} A) \right|, \tag{9}$$

does in the thermodynamic limit. We note that the ETH in the strong sense is a sufficient condition for the weak ETH. The ETH in the weak sense allows an infinitesimal number of the eigenstates which give the EEVs far off its average in the window $\text{tr}(\rho_{E,\delta}^{\text{mic}} A)$. These states are called the rare states which are suggested to be seen in integrable systems.[13]

Analyzing these two in terms of the finite-size scaling is our study in the next section.

3.2. Other Two Scenarios

The ETH concerns the magnitude of the fluctuation of the EEV, A_n. On the other hand, we can seek other scenarios for thermalization in the correlation between $\{A_n\}_n$ and $\{|c_n|^2\}_n$. Since the number of the eigenstates in the energy window is exponentially large in N, Eqs. (6) and (7) can be equal if there is little correlation between $\{A_n\}_n$ and $\{|c_n|^2\}_n$.

3.2.1. Typicality argument

Let us assume that $\{|c_n|^2\}_n$ are generated by an identical distribution where a single constraint, $\sum_n |c_n|^2 = 1$, is imposed. If the number of the eigenstates is large, the statistical average of $\text{tr}(\rho_{\text{DE}} A)$ is equal to $\text{tr}(\rho_{E,\delta}^{\text{mic}} A)$

in the leading order of $\mathcal{N}_{E,\delta}$. The variance turns out to be $\mathcal{N}_{E,\delta}^{-1}\sigma_A^2$. This means that the difference between $\text{tr}(\rho_{\text{DE}}A)$ and $\text{tr}(\rho_{E,\delta}^{\text{mic}}A)$ is of the order of $\mathcal{N}_{E,\delta}^{-1/2}\sigma_A$ for a typical initial state. Namely, when we have no special reasons why $\{A_n\}_n$ and $\{|c_n|^2\}_n$ are strongly correlated, we can expect the difference to be

$$\left|\text{tr}(\rho_{\text{DE}}A) - \text{tr}(\rho_{E,\delta}^{\text{mic}}A)\right| \leq C\mathcal{N}_{E,\delta}^{-1/2}\sigma_A, \tag{10}$$

where C is a constant independent of $\mathcal{N}_{E,\delta}$.

Since $\mathcal{N}_{E,\delta}$ is exponentially large, the second factor on the right-hand side assures that the left-hand side is small enough even if σ_A does not converge to zero.

3.2.2. Eigenstate Randomization Hypothesis (ERH)

The eigenstate randomization hypothesis (ERH) states that the EEV, A_n, fluctuates "randomly" in the microcanonical window.[14] While the ETH concerns the magnitude of the fluctuation of the EEV, the ERH does the internal structure of it. If $|c_n|^2$ smoothly varies in the microcanonical window, the ERH gives an extra convergence factor on the difference between $\text{tr}(\rho_{\text{DE}}A)$ and $\text{tr}(\rho_{E,\delta}^{\text{mic}}A)$ as the typicality argument does.

We define the sequence of $\{A_n\}_n$ to be "random" in the following way. For convenience, we relabel n so that $n = 0, 1, \ldots, \mathcal{N}_{E,\delta} - 1$ represent the eigenstates in the microcanonical window. We assume that the support of $|c_n|^2$ is included in $[E - \delta, E + \delta]$. We define a set of sequences labeled by an integer m which are denoted by $\{B_k^{(m)}\}_k$ as follows:

$$B_k^{(m)} \equiv \frac{1}{m}\sum_{\alpha=0}^{m-1} A_{km+\alpha} \quad (k = 0, 1, \ldots, M - 2), \tag{11}$$

$$B_{M-1}^{(m)} \equiv \frac{1}{n}\sum_{\alpha=0}^{n-1} A_{k(M-1)+\alpha}, \tag{12}$$

where M is the smallest integer not less than $\mathcal{N}_{E,\delta}/m$ and $n \equiv \mathcal{N}_{E,\delta} - kM$. We call the sequence the m-th coarse-grained sequence. The standard variance of the m-th coarse-grained sequence depends on m and is denoted by $\sigma(m)$. If $\sigma(m)$ decays in a power law in m for $m \ll \mathcal{N}_{E,\delta}$, or there exists a positive number γ satisfying $\sigma(m) = \sigma_A m^{-\gamma}$, the original sequence, $\{A_n\}_n$, is defined to be "random".

We define $|c_n|^2$ to be smooth if there exists a positive real number D of the order of $\mathcal{N}_{E,\delta}^0$ which satisfies

$$\left| \, |c_{n+1}|^2 - |c_n|^2 \, \right| \leq \frac{D}{\mathcal{N}_{E,\delta}^2}. \tag{13}$$

This happens, for example, when the initial state is prepared by exciting the system in the ground state with a laser whose spectrum is smooth.

If A_n is random and $|c_n|^2$ is smooth, we can show the following inequality:[14]

$$\left| \mathrm{tr}(\rho_{\mathrm{DE}} A) - \mathrm{tr}(\rho_{E,\delta}^{\mathrm{mic}} A) \right| \leq D' \mathcal{N}_{E,\delta}^{-\frac{\gamma}{\gamma+1}} \sigma_A, \tag{14}$$

where $D' \equiv \gamma^{1/(\gamma+1)}(1 + \gamma^{-1})D$.

Again, since $\mathcal{N}_{E,\delta}$ is exponentially large, the second factor on the right-hand side assures that the left-hand side is small enough even if σ_A does not converge to zero.

4. Finite-Size Scaling Analysis of the ETH

We investigate the ETH in terms of the finite-size scaling in the Lieb-Liniger model which is a model for a one-dimensional interacting Bose gas.[15] Since the many-body eigenstates of this model are obtained analytically with the Bethe ansatz, we can avoid the computational complexity for diagonalizing the Hamiltonian and address larger system sizes.

The Hamiltonian of the Lieb-Liniger model is[16]

$$H = \int_0^L \mathrm{d}x \left[\partial_x \Psi^\dagger(x) \partial_x \Psi(x) + c \Psi^\dagger(x) \Psi^\dagger(x) \Psi(x) \Psi(x) \right], \tag{15}$$

where Ψ $(\Psi^\dagger)$ is the annihilation (creation) operator for a Boson, L is the linear dimension of the system, c is the strength of of the contact interaction, and we set $\hbar = 2m = 1$. The periodic boundary condition is imposed.

We calculate the real and imaginary parts of the eigenstate expectation values for $\Psi^\dagger(x)\Psi(0)$ for $x = L/2N$ and $c = 10$ where N is the number of Bosons. These correspond to the EEVs for the Hermitian operators $\frac{1}{2}\{\Psi^\dagger(x), \Psi(0)\}$ and $\frac{1}{2i}[\Psi^\dagger(x), \Psi(0)]$ where $\{A, B\} \equiv AB + BA$ and $[A, B] \equiv AB - BA$. For N ranging from 5 to 32 and with N/L fixed to be unity, all the EEVs are calculated for the eigenenergies less than $E_{\mathrm{g}} + 10$ where E_{g} is the energy of the ground state. The real and imaginary parts are plotted in Fig. 1.[15] They show systematic linear behaviors and the data lie around them.

We conduct linear fittings of the data for each N and subtract them from the original data to analyze the deviation of the EEV's from the local

268

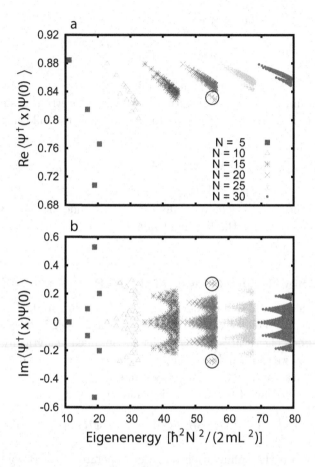

Fig. 1.   The real (a) and imaginary (b) parts of the expectation values of $\Psi^\dagger(x)\Psi(0)$ over each eigenstate plotted against the corresponding eigenenergy. All the eigenstates whose eigenenergies are between E_g and $E_g + 10$ are plotted for $N = 5$ (filled square), 10 (triangle), 15 (asterisk), 20 (cross), 25 (plus), and 30 (dot). The circled data in $N = 20$ show the emergences of other branches.

averages. The maximum and the variance of the subtracted data are shown in Fig. 2.[15]

The left panel in Fig. 2 shows that the ETH does not hold in the strong sense. There appear jump-ups from between $N = 15$ and 16 due to the appearances of new branches of data circled in Fig. 1.

The right panel in Fig. 2 shows that the ETH does hold in the weak sense and the standard variance decays in a power law in N, $\sigma \propto N^{-\alpha}$,

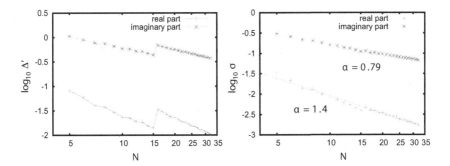

Fig. 2. The maximum deviation Δ' (left) and the variance σ (right) around the systematic linear behaviors in Fig. 1 plotted against the number of particles, N. The former does not converge to zero due to the jump-ups seen between $N = 15$ and 16 originated from the appearance of new branches of data circled in Fig. 1. The latter converges to zero in a power law in N.

where α turns out to be 1.4 and 0.79 for the real and imaginary parts, respectively.

While the ETH seemed not to hold in integrable systems in the previous studies, we have seen that the ETH can hold in the weak sense.

5. Discussion

The power law scaling of the variance of the EEVs, $\sigma \propto N^{-\alpha}$, enables us to discuss the relative importance between the ETH and the other two scenarios. If either of the typicality argument or the ERH works, it gives the convergence factor consists of $\mathcal{N}_{E,\delta}$, which is exponentially large in terms of N: $\mathcal{N}_{E,\delta} = e^{O(N)}$. On the other hand, the ETH in the weak sense gives the convergence factor, $\sigma_A \propto N^{-\alpha}$ (see Eqs. (10) and (14)). This shows that the ETH contribution to thermalization is only a logarithmic correction to the other two scenarios.

6. Prospect

The finite-size scaling analysis is being studied for small nonintegrable systems.[17] This suggests that the convergence of the variance of the EEVs is much faster than that in the Lieb-Liniger model. It is important to conduct the finite-size scaling analysis in nonintegrable systems up to larger system sizes but challenging due to the numerical complexity for diagonalizing Hamiltonians.

Acknowledgments

This work was supported by KAKENHI Grant No. 22340114 from the Japan Society for the Promotion of Science, and a Grant-in-Aid for Scientific Research on Innovation Areas "Topological Quantum Phenomena" (KAKENHI Grant No. 22103005), and the Photon Frontier Network Program from MEXT of Japan. I also acknowledge support from JSPS (Grant No. 248408).

References

1. A. Polkovnikov, K. Sengupta, A. Silva and M. Vengalattore, *Rev. Mod. Phys.* **83**, 863(Aug 2011).
2. P. Reimann, *Phys. Rev. Lett.* **101**, p. 190403(Nov 2008).
3. M. Rigol, V. Dunjko and M. Olshanii, *Nature* **452**, 854(Apr 2008).
4. S. Sugiura and A. Shimizu, *Phys. Rev. Lett.* **108**, p. 240401(Jun 2012).
5. S. Sugiura and A. Shimizu, *Phys. Rev. Lett.* **111**, p. 010401(Jul 2013).
6. H. Tasaki, *Phys. Rev. Lett.* **80**, 1373(Feb 1998).
7. A. J. Short, *New Journal of Physics* **13**, p. 053009 (2011).
8. J. M. Deutsch, *Phys. Rev. A* **43**, 2046(Feb 1991).
9. M. Srednicki, *Phys. Rev. E* **50**, 888(Aug 1994).
10. M. Rigol and L. F. Santos, *Phys. Rev. A* **82**, p. 011604(Jul 2010).
11. L. F. Santos and M. Rigol, *Phys. Rev. E* **81**, p. 036206(Mar 2010).
12. A. Pal and D. A. Huse, *Phys. Rev. B* **82**, p. 174411(Nov 2010).
13. G. Biroli, C. Kollath and A. M. Läuchli, *Phys. Rev. Lett.* **105**, p. 250401(Dec 2010).
14. T. N. Ikeda, Y. Watanabe and M. Ueda, *Phys. Rev. E* **84**, p. 021130(Aug 2011).
15. T. N. Ikeda, Y. Watanabe and M. Ueda, *Phys. Rev. E* **87**, p. 012125(Jan 2013).
16. E. H. Lieb and W. Liniger, *Phys. Rev.* **130**, 1605(May 1963).
17. W. Beugeling, R. Moessner and M. Haque, *ArXiv e-prints* (August 2013).

SECOND LAW OF THERMODYNAMICS WITH
QC-MUTUAL INFORMATION

TAKAHIRO SAGAWA[†]

Department of Basic Science, The University of Tokyo, Bldg.16
#727A, Komaba 3-8-1, Meguro-ku, Tokyo, 153-8902, Japan

I'd like to discuss generalizations of the second law of thermodynamics in the presence of quantum information processing such as quantum measurement and quantum feedback control.[1,2] A quantum information content, referred to as QC-mutual information, is shown to play a crucial role. We also discuss the generalization of a quantum generalization of the Hatano-Sasa inequality for transitions between nonequilibrium steady states with quantum feedback control.

References

1. T. Sagawa and M. Ueda, Phys. Rev. Lett. **100**, 080403 (2008).
2. T. Sagawa and M. Ueda, Phys. Rev. Lett. **102**, 250602 (2009); **106**, 189901(E) (2011).
3. T. Sagawa, arXiv:1202.0983 (2012); as a chapter of: M. Nakahara and S. Tanaka (eds.), "Lectures on Quantum Computing, Thermodynamics and Statistical Physics", Kinki University Series on Quantum Computing (World Scientific, 2012).